PLUTO AND CHARON

PLUTO AND CHARON

Ice Worlds on the Ragged Edge of the Solar System

ALAN STERN

JACQUELINE MITTON

A Wiley-Interscience Publication

JOHN WILEY & SONS, INC.

New York • Chichester • Weinheim • Brisbane • Singapore • Toronto

This text is printed on acid-free paper ♾.

Published simultaneously in Canada.

For ordering and customer service, call 1-800-CALL-WILEY.

Library of Congress Cataloging in Publication Data:

Stern, Alan, 1957–
Pluto and Charon: Ice Worlds on the Ragged Edge of the Solar System / by Alan Stern, Jacqueline Mitton.
p. cm.
Includes bibliographical references and index.
ISBN 0-471-15297-8 (cloth : alk. paper)
ISBN 0-471-35384-1 (paper)
1. Atomic absorption spectroscopy. 2. Environmental monitoring.
I. Pluto (Planet) 2. Charon (Satellite) I. Mitton, Jacqueline.
II. Title.
QB701.S84 1998
523.48′2—dc21 96-9502
CIP

Printed in the United States of America

10 9 8 7 6 5 4 3 2

To Sarah and Kate,

Sisters for a new century.

—Alan Stern

CONTENTS

PREFACE

This book was born in the summer of 1994, when Jacqueline Mitton interviewed Alan Stern about Pluto for a BBC radio program. The two of us immediately hit it off, and began exploring how we might pursue a proposal for a book about Pluto. Alan had been contemplating an idea for such a book for some years. With Jacqueline's experience of writing and editing astronomy books, and Alan's intimate knowledge of Pluto, we thought we'd have a good chance of putting together something that would be entertaining, engaging, and fully up-to-date. After all—no book on Pluto had been written since 1980—the veritable dark ages of Pluto studies.

We developed a book proposal together, and it was accepted by John Wiley & Sons in late 1995. We then divided the work. Alan did the writing; Jacqueline provided critiques of the text as the chapters were completed, and she took on the enormous job of creating and gathering illustrations. The text was drafted and the illustrations collected during the first half of 1996. The material was reviewed and revised during the remainder of 1996. To ensure accuracy and balanced treatment, we asked various scientific experts and lay people to read different chapters or pairs of chapters. Every chapter benefited from at least four such critiques. Additionally, the whole text was read by John Wiley & Sons' technical referees, Marc Buie and Steve Shore, who were both tremendously helpful. We are indebted to all of these readers.

One thing that was difficult for us was figuring out a method of handling the Pluto research in which Alan himself had been involved. Because there are two of us, it wasn't right to use the first person, "I," nor was it correct to say "we." It also didn't feel right to us or to our publisher to refer to this work as if it were someone else's—that is,

"Stern's work." So what did we do? We simply referred to the work without direct attribution. If you are curious about which Pluto research discussed in this book that Alan was involved in, the primary ones were the question and resolution of why Pluto is bright, the measurement of Pluto's brightness temperature at Institute Radio Astronomie Millimetrique (IRAM) in 1993, the search for small satellites around Pluto, the exploration of Pluto's dynamics with Hal Levison, the "Thousand Plutos" concept, some of the recent modeling work on the origin and evolution of the Kuiper Belt, and the Hubble Space Telescope (HST) imaging and mapping project for Pluto.

Our goal has been to write an up-to-date, but rather different kind of a book about Pluto. We wanted to create something that would be useful to students of astronomy and planetary science but would also be interesting for the layperson. In fact, it is our hope that the majority of our readers will be astronomy enthusiasts who see Pluto as we do—as a paradigm for the modern approach to exploring the planets.

We wanted to thread three themes through the writing. The first is the story of the advances made possible by the recent technological explosion in groundbased astronomical instrumentation. Since the late 1970s, new instruments (CCD imagers, infrared spectrographs, and spacecraft observatories) have dramatically increased the grasp of the telescopes with which the planets can be explored from Earth and from Earth orbit by a factor of a thousand. The impact of such devices on planetary astronomy is best illustrated by their use on Pluto, precisely because Pluto only became a subject of serious study as a result of this revolution in astronomical instrumentation.

The second theme is the revolution in scientific perspective that spacecraft studies have brought. From the first *Mariner* flyby of Venus in 1962 to *Voyager*'s last encounter, with Neptune in 1989, planetary scientists have, time upon time, seen hard-won preconceptions proven embarrassingly simplistic—and sometimes just flat wrong—when spacecraft arrive. In hindsight, the reason for this is obvious: Spacecraft observations achieve a resolution and clarity that just isn't approachable from the ground, or even from Earth orbit. The tension between widened scientific perspectives and the history of failed scientific expectations makes one wonder whether all of our efforts can get Pluto, the last astronomer's planet, right—from afar.

Finally, we wanted to involve the issue of how humans, born and bred on Earth, can come to know exotic and far distant worlds as real places. It is a question of how the mind's eye works. The kings and queens of 16th-century Europe once imagined the Spice Islands, China, and North America. Our generation's Spice Islands are the worlds of the solar system.

If this book succeeds in reaching our goals and in aptly addressing these things, it is in large measure because of the many friends and colleagues who provided advice, critique, practical help, and encouragement along the way. That list includes Fran Bagenal, Kelly Beatty, Rick Binzel, Bonnie Buratti, Steven Brewster, Marc Buie, Steven Dick, Jim Elliot, Robert Farquhar, Don Hassler, Helen Horstman, Garth Hull, Jan Ishee, Hal Levison, Jonathan Lunine, Bob Marcialis, Bill McKinnon, Dianna Miller, Robert Millis, Simon Mitton, Tony Reichardt, Steve Shore, Kerry Soileau, John Spencer, Rob Staehle, Carole Stern, Sarah Stern, Mark Sykes, Clyde Tombaugh, Roderick Willstrop and the Library of the Institute of Astronomy at Cambridge (UK), Eliot Young, and Lesley Young. Alan also wishes to thank Larry Trafton and Harlan Smith for interesting him in Pluto in the first place, back in 1979 and 1980.

We are also grateful to Greg Franklin of John Wiley & Sons for his constant encouragement, practical support, and patience while we were working on this book, and to Andrew Prince for his dedicated and marvelous editing.

The Pluto–Charon system is a fascinating frontier for astronomy, for astronomers, and for the human spirit. We hope very much that NASA doesn't lose sight of Pluto's scientific value and fascination for the public as it plans its explorations!

Alan Stern and Jacqueline Mitton

Spring, 1997

PROLOGUE: !ENCOUNTER

After traveling almost 10 years and more than 3 billion miles to cross the breadth of the solar system, the speeding spacecraft is almost upon its twinned target. The years of anticipation are almost over. There is no time to waste.

If you could be there—there, alongside the first robotic explorer to reach the last of the nine planets—you would find yourself in a strange but somehow vaguely familiar setting. The vacuum of space on the outer fringes of the planetary system is much as it is at home, between the Earth and the Moon: empty, odorless, and crystal clear. So too, the stars still shine steadily as they do in Earth orbit, and in their familiar constellations as well.

But everything is not the same. The Sun is so distant that it barely shows any disk at all, and the Sun is so feeble that its light is diminished 1,000 times—to an eternal, almost maddening dusk that is more akin to a snowy moonlit night than any daytime you ever knew. In turn, because the sunlight is so thinned out, it cannot warm very well. As a result, even the blackest, most sunlight-absorbing surfaces can barely reach 300 degrees below 0°F. This wilderness is a place beyond human experience. Fortunately, the sleek but silent probe named Pluto Express *is right at home here. This is the place it was designed for—and the day, too. Its time to explore new worlds is about to arrive.*

Looking sunward, whole worlds appear as specks, mere flotsam against the emptiness we so simply call "space"; their faint crescent disks are only dim embers, lost against the Sun's glare.

Looking ahead, however, there is something different. There, against the constellation Taurus the Bull is the binary planet Pluto–Charon, moving slowly against the stars.

Although Pluto is almost 1,600 miles in diameter, the half a million miles that still lie between the spacecraft and this frozen, ancient, ice-laden world reduce it to the size of a pinkish little marble held at arm's length. From this vantage point, Charon, about half the size of Pluto, is just two degrees (about the width of your palm at arm's length) to the right of Pluto. In contrast to Pluto's warmer color, Charon presents an almost colorless, ashen reflection of the feeble yellow sunlight that falls on its surface.

The Pluto–Charon system is remarkable among the planets. It is the only real example of a double-planet system ever found. Further, Pluto itself is the only planet with an atmosphere that forms and decays each orbital cycle, and it loses its atmosphere to space at a far faster rate than does any other planet, as if it were a comet on a planetary scale. Pluto also is apparently the rockiest planetary body in the outer solar system, yet it contains a substantial supply of the lightest common icy substance found in the deep outer solar system—methane. And amazingly, the snows at the feet of Pluto's mountains and on the floors of its valleys primarily consist of some of the same stuff you inhale with every breath—nitrogen.

Most remarkably, if the modern theory of Charon's origin is correct, then the Pluto–Charon binary is also the best analog we have for understanding the formation of the Earth–Moon pair.

Even more dramatically, however, it has become clear in the 1990s that Pluto and Charon are not odd misfits among the giant planets, but are instead the first and most easily detected examples of a large population of small, ice-dwarf planets that were formed in the ancient outer solar system. As such, Pluto and Charon represent more than just a frontier: They are now understood to be a fundamental link, and the gateway, between the classical planets and the myriad small bodies orbiting the Sun beyond Neptune. Among the five fascinating and fantastic frozen outer planets, Pluto is the *rogue—and only Pluto remains unexplored.*

Now, after a journey of almost a decade, the first of two Pluto Express *spacecraft is almost upon its goal, and the anticipation is palpable. The robot's very human masters pace—and they worry. They try to help the probe along, but they know that the events they now monitor took place almost 5 hours before. Were word of a problem to reach the warm Earth 3 billion miles sunward, it would be too late to do anything about it. The spacecraft are on their own.*

The twin, dartlike, Pluto Express *spacecraft were born from the ashes of an earlier era of planetary exploration, when NASA once sent houseboat-sized probes to the outer planets. The slow and heavy precursors to the* Pluto Express *weighed in at several thousand pounds each. So heavy were the old explorers that even the largest launch vehicles could not shove them hard enough to make the crossings to Jupiter and Saturn in less than six or eight years. By contrast, lithe* Pluto Express *flashed past Jupiter's orbit in half that time. Good thing, too: The distance from Earth to Jupiter is barely more than 10% of the trip to Pluto.*

Though spare, each 250-pound craft also packs more sophistication within its skins than its successful but Sumo-like predecessors. Aboard the Plutocraft are delicate thrusters for precise pointing and nimble course control, computers possessing state-of-the-art autonomous, "thinking" (artificial intelligence) software, and a telecommunications system capable of receiving the faint, diluted signals sent to direct it from Earth. Even more impressively, within its belly, each craft packs imaging cameras, spectrometers, and an ultrastable radio beacon; this payload is capable of probing Pluto and Charon with 10 times the precision and perhaps twice the resolution that the massive Galileo *spacecraft probed Jupiter in the late 1990s. They say amazing things come in small packages. Indeed.*

Speed is the key that unlocks the universe, and for the Pluto Express, *speed was a gift given to it by its launch vehicle, which, fully assembled, stood almost 200 feet high. Little* Pluto Express *looked like a mere pin atop the tall white javelin that set it coursing across the great expanse of our solar system.*

Then, after the rocket's mad push away from Earth, there was the journey: 3000 days through the endless, eternal vacuum that knows no ebb and flow, no seasons, no emotion—only silent running—past Venus to gain speed, past Earth, then across the orbit of Mars, through the asteroidal shards that remain of what might have become the fifth planet, past Jupiter to gain yet more speed, and then off into the darkened, wild abyss where Saturn, Uranus, and Neptune reign.

Speeding, cruising, without sound (or even whisper), the twin Pluto Express *spacecraft made the great crossing. And as they arced out of the planetary system, the constant tug of the Sun's gravity made the journey to Pluto a long, ever-slowing uphill glide.*

Time passed. So did presidents, and styles, and hot stocks, and cool trends. Earth raced round the Sun ten times and headed toward her 5-billionth birthday.

In California, Colorado, and Texas, and in Germany and Russia, too, the builders of these two craft waited. Years passed. Their children grew tall—and still the builders waited. And, as is the nature of journeys across the Great Deep of interplanetary space: Nothing much happened, even at 35,000 or 40,000 miles per hour. Why? Because the solar system is enormous beyond human comprehension. We can say we understand the scale—billions of miles, but does it have meaning? The numbers are so sterile, so alien—so surreal.

So, despite the numbing speed, the trip was timeless. Astronaut Alan Bean once remarked that the strangest part of his 3-day dash across the 240,000-mile gap between the Earth and the Moon was "that you never pass anything." Imagine the 6-billion-kilometer crossing to Pluto!

Oh, along the way to Pluto there was the occasional hit by a microscopic dust grain adrift from some long-past comet. And there were the annual status checks and the every-so-often course-correction burns directed from Earth, "below." But basically, nothing happened. Patience is a virtue in space exploration.

Now, however, the trip is almost over. The first Pluto Express *spacecraft is approaching its target at almost 10 miles per second, some 30 times faster than the fastest supersonic fighter ever flew. As the first craft is nearing its long-sought target, its sister follows 200 million miles behind, in order to make a second, more refined reconnaissance a few months hence.*

Aperch the leading spaceship, now only hours from its goal, Pluto appears perhaps half the size of the Moon on a midsummer night. Sharp provinces with subtle shades of yellow and ruddy pink abound on its surface, and a thin haze lies over its rounded limb. The telescopes aboard the craft can already see craters, mountains, and long, linear ridges on this outpost of the cryogenic outer limits. Even Charon is beginning to look like a globe; and something is glinting on its morning limb . . . .

Until a spacecraft actually reaches Pluto, we will never know what that something might be, glinting on Charon's sunlit limb. Indeed, as we write this book, such a scenario remains only a dream.

Still, in the pages that follow, we can explore the Pluto–Charon binary to the extent possible from the vantage point of late 1990 astronomy, and the information gleaned over more than 65 years of astronomical research since Pluto's discovery.

To do so, however, we must first travel in a different direction. Rather than upward, toward the worlds beyond the sky, we must first travel backward in time, toward the dawn of the 20th century. So we begin with a place that is much more remote from us than Pluto, and far less accessible—on a frosty Arizona hilltop, in February of 1930. . . .

CHAPTER

1

NEW FRONTIER

Ever since celestial mechanics in the skillful hands of Leverrier and Adams led to the world-amazed discovery of Neptune, a [belief] has existed begotten of that success that still other planets lay beyond, only waiting to be found.

PERCIVAL LOWELL, 1915

A PLANET HUNT

The faint starry image jumped a little—very little, in fact. But the heart of young Clyde W. Tombaugh (Figure 1.1), who saw it twitch late that Tuesday afternoon in February of 1930, jumped a good bit more.

Tombaugh had been hired by Lowell Observatory, in Flagstaff, Arizona, about a year before. His job: to take up the reins of long-dead Percival Lowell's quest for that romantic denizen of the deep outer solar system—Planet X. Now, barely a year later, 24-year-old Tombaugh had found it.

Who was this cherubic planet finder? He was a midwestern American, born in Streator, Illinois, in 1906, the eldest of his parents' six children. Clyde (1906–1997) had become interested in astronomy early in his childhood, and this boyhood interest was magnified through the loan of a 3-inch (7-cm) telescope[1] from his uncle Lee. By the time his father Muron moved the family to a rented wheat and corn farm in Burdett, Kansas, in mid-1922, Clyde was already an avid amateur. In high school, Clyde played track and field, "dabbled in Latin," and

[1] *Telescope size* refers to the diameter of the telescope's main lens or mirror, which determines how much light the telescope can gather.

FIGURE *1.1*

Young Clyde Tombaugh, entering the dome of the 13-inch telescope at the Lowell Observatory in the early 1930s: He is carrying one of the large photographic plate holders for the "Plutoquest telescope." (Lowell Observatory photograph)

played football with his friends on weekend afternoons. His passion, however, was astronomy, and his classmates knew it. In fact, his fellow Burdett High School seniors wrote in their 1925 yearbook that Clyde had his head "in the stars."

Too poor to afford college, Tombaugh took the job at Lowell equipped with little besides a high school education and a fire inside himself to become a *working* astronomer.

The job at Lowell was young Tombaugh's first experience away from his family. It was a little scary to go so far from home. There, away from friends and family, Clyde was to laboriously photograph the sky and inspect the images in search of astronomer–aristocrat Percival Lowell's (Figure 1.2) last, some thought chimerical, obsession, this Planet X. He didn't care much if it was tough, almost thankless work. Never mind that there wasn't any assurance of ultimate success. He just dove right in. Night after night. Week after week. Month after month. But his labor *did* pay off—for on that cold February afternoon in Flagstaff, Arizona, Clyde Tombaugh found one of the astronomical prizes of the young century.

The decades-long search for Planet X had been driven by two motivations—one scientific, one more instinctive. First, there had been the mounting observational evidence in the early part of the century that some unseen object was tugging at Uranus and Neptune, causing its course on the sky to differ from predictions. Second, there was the sheer lure of it—the attraction of finding a new world, of making a mark on the immortal annals of discovery, of tilting at Quixote's old windmill.

The search for the ninth planet was largely initiated by Percival Lowell, a wealthy and scientifically literate aristocrat born just before the U.S. Civil War. It's a good thing that an instinctive motivation to make the search was so strong in Lowell, because he didn't know that the measurements (made by others, mind you) indicating the positions of Uranus and Neptune were slightly off predicted course, were wrong. Uranus and Neptune were right on course.

FIGURE *1.2*

Percival Lowell (1855–1916), founder of the Lowell Observatory at Flagstaff, Arizona, was the chief motivator behind the observatory's original search for Planet X; ca 1910. (Lowell Observatory photograph)

Oftentimes, ignorance is bliss. And ignorance of the true situation caused Lowell to use the incorrect "residuals"—the difference between the expected course of Uranus and Neptune, and what had been measured—to calculate where in the sky the phantom planet should be.

As early as 1905, Lowell organized the first extensive search for his elusive prey, which he called "Planet X." This first search was not fruitful. But Lowell was not alone in planet hunting. Several other astronomers also pursued the same goal, including MIT-educated astronomer William H. Pickering. Pickering called the phantom "Planet O," and organized several searches for it in the early years of the century.

All together, at least eight searches were undertaken by Pickering and Lowell between 1905 and 1920. In these searches more than 1000 photographic plates[2] were made—all to no avail. Those were long years of depressingly ill-rewarded work. Almost two decades of it, in fact, during which four U.S. Presidents passed the American stage, automobiles exploded into the commonplace, the motion-picture industry was born, and there came (and thankfully went) The Great War. As the years dragged by, Percival Lowell grew old. Saddened by the World War, exhausted, and discouraged, Lowell suffered a stroke and passed away in November 1916.

When Lowell died, his trust ensured the survival of his observatory and its staff in Flagstaff. "Percy's" will stated that the search for X should continue, and so it did. His dedicated assistants, led by the Observatory's acting director (Figure 1.3), Vesto M. Slipher, took up the task. But just in case the faint-hearted might someday hence waver in their search, Percy had himself interred in an ornate blue-and-white marble mausoleum less than 50 meters from the main administration building of his former observatory (see Figure 1.4). It was just a little reminder, for the staff.

Entombed reminder or not, Lowell's search for X *did* cool after Percy's death. However, in California, Pickering and his assistant Milton Humason (a man who was to later become one of the most widely respected midcentury astronomical observers) undertook an additional search in 1919, but once again—nothing was found.

Meanwhile in Arizona, Lowell's widow sued his estate for a significant part of the million dollars he had intended for the observatory.

---

[2]In those days, astronomical photographs were made using glass plates coated with light-sensitive chemical emulsions.

FIGURE 1.3

Vesto Slipher (1875–1969) in 1911: After Lowell died in 1916, Slipher became acting director of the Lowell Observatory. Ten years later, he was confirmed as director. In 1929, Slipher hired Clyde Tombaugh as an assistant to help in the search for Lowell's predicted ninth planet, then called "Planet X." (Lowell Observatory photograph)

Litigation fees drained the operating funds of the observatory, and despite the efforts of Slipher and other astronomers at Lowell Observatory, the search for X stalled. In fact, almost a decade passed.

Then, in 1925, Percy's brother Abbott contributed the funds to commission a new tool for hunting Percival's suspected planet. That tool was a 13-inch (33-centimeter diameter) refracting telescope and camera that combined to create an unusually large and useful field of view for planet hunting—164 square degrees.[3] With its combination of broad search swath and good sensitivity, the new telescope (Figure 1.5) would make it worthwhile to renew the search.

By 1928, when the new telescope's arrival was imminent, Vesto Slipher himself had grown a little older, and his responsibilities had expanded to include the full directorship of the observatory. So Slipher

[3]A square degree is a rectangular patch of sky one degree on a side.

FIGURE 1.4

A recent view of the mausoleum at the Lowell Observatory where the remains of Percival Lowell are interred. (Photo by Richard Oliver; all rights reserved)

decided to hire a technician to assist in the laborious planet search, and he had just the man in mind.

Slipher offered the job to a young man who had been enthusiastically corresponding with him and the other astronomers at Lowell for some time. The young man wrote time after time to share the careful sketches of Mars and Jupiter that he had made using a small telescope that he himself had built. The young man's talent and chutzpah were remembered and landed him the job. That young man was Clyde Tombaugh.

"YOUNG MAN, I AM AFRAID YOU ARE WASTING YOUR TIME."

Tombaugh arrived in Flagstaff from Kansas by train on the wintry afternoon of January 15, 1929. At the station, a graying Vesto Slipher met Clyde Tombaugh for the first time. Clyde carried a single, heavy trunk, which was laden more heavily with books than clothes. In the letter that offered Clyde the job at Lowell, Slipher hadn't told Tombaugh that he was being brought to Flagstaff to search for Planet X, merely that he would be involved in "photographic work." It was only on his arrival that Tombaugh realized what his job would center on.

His first duties? The new planet-finder telescope still wasn't quite

FIGURE *1.5*

The 13-inch Lawrence Lowell telescope (named after Percival's brother), with which Tombaugh discovered Pluto, as it was originally in its dome at the Lowell Observatory on Mars Hill, Flagstaff. In 1970, the telescope was moved to the Lowell Observatory's site at Anderson Mesa, 12 miles southeast of Flagstaff, but in 1995, it was restored to working order at its original site on Mars Hill. (Lowell Observatory photograph)

ready. So Clyde, astronomer in waiting, was assigned tasks ranging from showing visitors around the observatory to shoveling snow and stoking the main building's furnace. Mercifully, however, by early April, the new telescope was finally ready, and Tombaugh could begin.

Initially, Slipher tutored Tombaugh in the observing technique: X was expected to be at least 15 times fainter than Neptune. They would have to make long, deep exposures of the sky so that the photographic plates could soak up enough light to detect the prospective planet. Because planets move perceptibly from night to night but stars do not, the key would be to spot a faint, slowly moving target, against the backdrop of myriad stars and galaxies. To do this, each plot of sky would be rephotographed over the course of several nights. Each plate would have to be exposed for an hour or more to detect the faint phantom they were searching for, and Clyde would have to precisely adjust the telescope to keep pace with the turning sky throughout each exposure. To ensure the darkest possible sky as an aid to spotting the faint and distant wanderer, the images were to be made around the time of the new Moon.

According to Clyde's account of the period, after a few nights of lessons, Slipher said, "You're doing all right. You're on your own." From then on, Tombaugh did the photography. Within a few weeks, Slipher also gave Tombaugh the responsibility for inspecting the plates for evidence of the hoped-for planet. In effect, the search for Planet X now rested entirely with one very young man.

Tombaugh told us in 1996 that few people, even astronomers, "have any concept of the grim [plate comparison] job that V. M. Slipher gave to me." We agree: Today, no one would stand for such a job—a computer would make the comparisons. But the late 1920s were another time, laden with a wholly different set of technologies to accomplish things. It was a time when the ocean was crossed by steamship, phone connections were made by human operators moving cables across a board, and most iceboxes actually used ice, delivered from distant frozen lakes. So too, the manual "technology" for comparing Tombaugh's images was, by today's standards, frighteningly primitive.

After a few months of work, Clyde realized that the most efficient place in the sky to search for a distant planet would be around the spot where a distant planet's motion, as seen from Earth, would be both fastest and most noticeable. At this place on the sky, called the "opposition point"—a distant planet seems to move westward—that is, backward, against the stars.

Over the course of a year the opposition point[4] sweeps out a great circle on the celestial sphere which astronomers call the ecliptic. This great circle passes through all 12 traditional Zodiacal constellations, plus Ophiuchus. A second important advantage of working at the opposition point, Tombaugh realized, was that the amount of motion a faraway planet displays at the opposition point is inversely proportional to its distance from the Sun—that is, the farther away an object is orbiting around the Sun, the more slowly it appears to move. Thus, by working at the opposition point Tombaugh could easily distinguish a distant, slowly moving planet from the numerous, nearby asteroids that also peppered the plates but moved much more quickly across them (Figure 1.6). This strategy was a crucial advance because it made the detection of truly distant objects so much more straightforward.

---

[4]The opposition point gets it name from the fact that it is the point opposite the direction of the Sun.

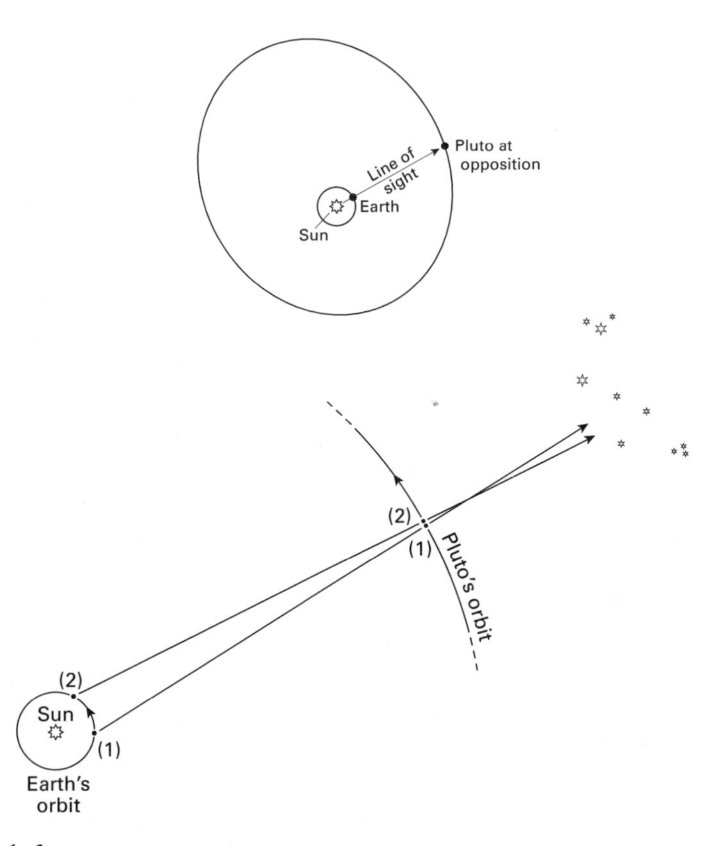

FIGURE 1.6

Top: A planet is "at opposition" when it lies in the part of the sky directly away from the Sun, as seen from Earth. This occurs when the Sun, the Earth, and the planet are in a line in space, as shown here schematically. This is the optimum time for making discovery observations for a number of reasons. One is that the planet reaches its highest point in the sky at midnight, so potential observing time is a maximum. It is also at one of its nearest points of approach to Earth. Another key factor is that a planet's distance can be directly measured at opposition by its rate of travel across the sky. Bottom: Earth travels farther than Pluto does in the time between two observations, (1) and (2). Traveling in its orbit six times faster than Pluto, Earth rapidly overtakes the more distant planet. As a result, Pluto's position in the sky changes relative to the starry background. The change in position over a given period of time is greatest near opposition. The stars themselves do not appear to change their positions between images taken a few days apart because they are thousands of times farther away than Pluto.

Tombaugh presented the opposition-point search plan to Slipher, who approved it. With this new technique in hand, Tombaugh went back to the telescope. Night after night, he imaged the sky, calculating that 30 plates would have to be taken each month to cover the opposition region of the ecliptic as it swept eastward. Tombaugh knew that he could only work during those nights each month when the Moon was new or nearly so.

When Tombaugh wasn't up nights photographing the opposition point, he was at work days, developing the plates and inspecting them for moving objects. It wasn't easy, it wasn't glamorous, and it wasn't even very interesting. Fortunately, Clyde's commitment was monumental. As was his concentration—which was well-needed in order to combat the sheer drudgery of methodically inspecting hundreds of plates, each of which contained 50,000 to 900,000 stars, looking to see whether one of the faint points of light on it might move a bit from night to night.

Clyde blinked (i.e., compared) his plates slowly, for hours on end, day after day (Figure 1.7). He said later that he had to take frequent breaks to clear his mind so he could keep concentrating. The penalty for missing the suspected prey was too great to permit his mind to become dulled by the tedium.

By the time he'd blinked a month's work of newly developed plates, the opposition point would have moved. So Clyde would photograph a new region, and develop the plates, and blink those too. And so on. And so on.

Oftentimes, the plates contained microscopic flaws that appeared to create a little jumping point of light when two plates were intercompared, as if to mock him—"Just testing." Almost every plate also contained a handful of asteroids, which sometimes gave false alarms. Fortunately, however, their motion at the opposition point was usually so large that Tombaugh could tell they were too close to Earth to be X itself. After all, he was looking for something that would move just a few millimeters between two plates; the asteroids moved 10 times that, or more.

In the course of his searches of the outer solar system, Tombaugh counted over 29,000 galaxies on his plates, plus 1,800 variable stars; and he discovered two new comets.

By February of 1930, Tombaugh had worked his way around the ecliptic to the constellation Gemini. He had made a few test plates in this region in early 1929 when the 13-inch telescope was first being set up, but at that time, he hadn't hit upon the idea of using the opposition point to make distant planet detections easier and more systematic.

FIGURE 1.7

Clyde Tombaugh in 1938 at the Zeiss blink comparator, which he used to examine pairs of photographic plates from the 13-inch telescope during his search for any object that moved against the background of stars: Following the discovery of Pluto, Tombaugh continued this dedicated but ultimately fruitless work, in an attempt to find planets beyond Pluto. The large glass plates Tombaugh used each measured 14 inches by 17 inches (36 by 43 centimeters). (Lowell Observatory photograph)

This time through Gemini, there would be no escape for the stealthy little world. *This time* Tombaugh ran right into his prey, and caught it.

THAT'S IT!

It was the 18th of February when Clyde examined his new plates of the star field around Delta Geminorum, the fourth brightest star in the constellation Gemini. There, a faint pinprick of light equivalent to a candle seen from a distance of 300 miles (480 kilometers) lost its 4-billion-year-old anonymity.

In the sky there are 15 million stars brighter than the pinprick Clyde spied hopping across a corner of the star fields of Gemini, but by blink-comparing the position of the anonymous little pinpoint (Figure 1.8) between the various January plates, he could see it move!

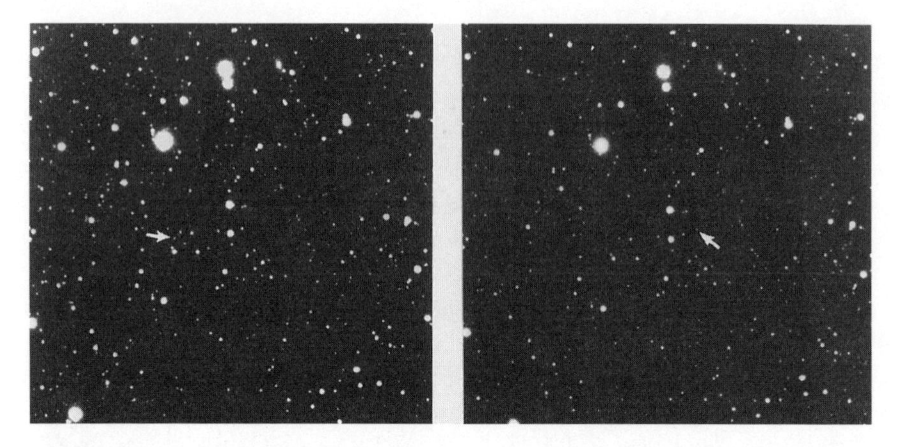

FIGURE 1.8

Portions of the original photographic plates on which Tombaugh first detected Pluto by noticing its change in position against the background star pattern: On each plate, the image of Pluto is arrowed. The left plate was taken on January 23, 1930, and the right one 6 days later, on January 29. (Lowell Observatory photograph)

It didn't jump far, only 3 or 4 millimeters, and the fact that the jump was so small was the exciting part, for that indicated that if the object was real—a big *if*, to be sure, then it surely lay beyond Neptune. "That's it!" he said to himself, but in his logbook, his very own "X" files, Tombaugh simply wrote, "planet suspect" and the coordinates of the tantalizing speck of light. It was 4 P.M.[5]

After he finished inspecting the Gemini plates, good soldier he, Tombaugh set out to confirm whether the faint little nomad was indeed real. The test he applied was deceptively simple, but stunningly powerful. Clyde had taken three other plates of the same region on the same nights, using the smaller, 5-inch-diameter telescope that he had bore-sighted onto the 13-inch as a kind of finder scope.

Tombaugh's planet suspect had to appear in the same place on these plates, as well, if it was real, rather than just some unlucky artifact of the 13-inch telescope and its camera. And it *was* there. Bingo! The little wanderer appeared right where it should on all three of these check plates. For nearly three-quarters of an hour, as he checked and cross-checked the moving pinpoint among the various plates, Clyde

[5]More than 65 years later, Tombaugh loved recalling how he discovered the ninth planet, "during the daytime!"

Tombaugh was the only person in the world to know that Planet X had very likely been found.

Then, trembling with excitement, Clyde went to see his boss, Slipher, and Lowell's assistant director, Carl Lampland. Standing at the door of Slipher's office, Tombaugh announced, "I have found your planet X." Clyde had never before come to make such a statement, nor set off any false alarm, and both men knew it. On hearing Tombaugh's words, Slipher stood up from his chair and, with Lampland, rushed to the blink comparator to check Tombaugh's plates. The two older men shortly confirmed Clyde's findings. Slipher's order: "Don't tell anyone until we follow it for a few weeks. This could be very hot news." Excited, but cautious, Vesto Slipher wanted more evidence.

Because 3 weeks had passed since the Gemini plates had been taken, an even more solid test presented itself. Based on the direction and degree of motion from January 21 to 29, a precise prediction of the planet's suspect's new location could be made. If the object was truly a distant world, it should show up on new plates about a centimeter from where it had been in late January. This kind of prediction check is a hallmark of observational astronomy. It had to be done—and done before any announcement could even be contemplated. It had to be done in part because it was in their own interest, but also, more importantly, because it was in their very bones as astronomers to check, and recheck, and find as many tests of their discovery as they could think of.

That night, however, clouds covered northern Arizona, which of course prevented planet hunting. Frustrated and bored, Tombaugh went into Flagstaff to catch a movie. *The Virginian*, with Gary Cooper, was showing. Word was it was going to be a hit.

After seeing the show, Tombaugh found the sky still cloud covered. The suspense must have been palpable, but Tombaugh had no choice: He would have to wait until the next night to try again.

Mercifully, the sky was clear on the 19th, and Clyde was able to capture another photograph of the little spot of light placed against the pregnant star field. By the next morning, Tombaugh had developed the new plate. Then he, Slipher, and Lampland examined it together—Slipher also brought his brother Earl along.

The scene must have been a classic, with four pairs of eyes jockeying for position at the blink comparator. And Lo and Behold—their little *It Girl* was there, exactly where the motion in the earlier images had predicted it would be! However, the object didn't show a round disk like other planets. Clyde's notebook records the reaction: "Each of the group took a look. There it was, a most unimportant-looking, dim, star-like object which had moved perceptibly . . . but no disk could be made out."

No disk? Every planet ever seen by human eyes had shown a disk. Like its motion, this was something that distinguished the appearance of a planet from the stars. . . . But maybe *this* one shouldn't show a disk. It was, after all, *very* faint. X had been predicted to be 10 or 20 times fainter than Neptune, but this thing (maybe they should have written it as *x*), was 250 times dimmer than Neptune.

Brother Earl reasoned that it might be small enough, and faint enough, and far enough away to just be a pinpoint, without detectable dimension in their tiny telescopes. To prove his point, Earl built a box with a disk-like hole in one side and a light within. He then took the box a mile from the observatory to convince the group that a faint and far away enough disk would look starlike. As Earl disappeared into the distance, the little disk grew smaller, and smaller, and smaller, until it finally became a pinpoint, just like X! It was a low-tech tour de force.

Now there was beginning to be the feeling that X was in fact the planet Percival Lowell had set out to discover so long before. Still, the astronomers wanted to be even more sure of their find. So, three weeks of checking and monitoring the motion of the new planet followed. Amazingly, during all this time, no one breathed a word of it beyond themselves and their immediate associates.

Then, finally, on the evening of March 12 in Flagstaff (but March 13 in Greenwich, England, where official time was kept), Slipher released the news. After having followed Pluto for seven full weeks, Slipher officially announced the discovery of the planet on behalf of Lowell Observatory.

Slipher had selected the official date of March 13 to make the announcement because it held a special significance: It was both the 149th anniversary of the date that Uranus, the first discovered planet, had been spotted in 1781, and it would also have been Percival Lowell's 75th birthday.

Slipher made the announcement on that date through two channels. First, a telegram was sent via Lowell Observatory's trustee (Percy's grandson, R. L. Putnam) to the Harvard College Observatory, which was the official clearinghouse for discoveries of new asteroids, comets, and novae. Telegrams were typically terse in those days, and relatively expensive. Slipher's read:

Systematic search begun years ago supplementing Lowell's investigations for Trans-Neptunian planet has revealed object which since seven weeks had in rate of motion and path consistently conformed to Trans-Neptunian body at approximate distance he assigned.

The honor of the second announcement was given the next morning to Lowell's widow, Constance. With Mrs. Lowell's announcement, the circle was complete. It was done.

By 1930, the global village was already a real (if comparatively quaint) construction. And via the media, the world learned that X existed—and the world responded. Headlines echoed the news: "Ninth Planet Found at Edge of Solar System"; "9th Planet, X, Is Found by U.S. Scientists." Clyde's parents found out from a reporter:

Within days, Clyde and company were famous. Today, the discovery would probably net the standard Warhol allotment of 15 minutes of fame, but thankfully, in 1930, fame lasted a little longer. The media rush following the discovery lasted for months.

PATHWAYS OF THE GODS

Something deep-seated in human nature calls on us to name things. It's almost as if a thing isn't real, or whole, until we name it—and so, X *had* to have a name.

Suggestions flooded in: "Zeus," "Cronos," "Lowell," "Minerva." Widow Lowell first liked "Zeus," but later suggested "Percival," then "Lowell," and then, finally, "Constance," her own name.

Dozens of other well-meaning suggestions came pouring in as well. Then there were hundreds, then thousands. But when all was said and done, the moniker for the newly discovered X that the Lowell staff preferred was the one suggested[6] by 11-year-old Venetia Burney of Oxford, England: *Pluto*—Pluto, the Greek god of the Underworld; the brother of Jupiter, Neptune, and Juno; and third son of Saturn, who was able, when he wished, to render himself invisible.

Both the American Astronomical Society and the UK's Royal Astronomical Society adopted Pluto as the official name and ♇ as the official symbol for the new world. ♇ was Percival Lowell's monogram.

While the international press heralded the discovery of Pluto, astronomers around the world rushed to work out the orbit of the new

---

[6]The French astronomer P. Reynaud had suggested Pluto as the natural mythic name for Lowell's putative Planet X back in 1919, but this was not remembered in 1930.

planet. This was the natural first order of business. Without an orbit, the ninth planet's ever-changing position could not be predicted, and the new world could become lost. Worse, its context among its eight familiar sisters could not be even crudely judged until its orbit was known.

What exactly is a planetary orbit? The orbits of the planets are, simply put, their tracks around the Sun. Each orbit a planet makes is a lap on this repetitive course, and one such lap requires a period called the planet's "year." For over four centuries, astronomers had developed an increasingly sophisticated understanding of planetary orbits.

The basic framework for understanding orbits was laid down in the 16th century by Renaissance astronomer Johannes Kepler. Kepler's discoveries resulted from the mental sweat equity he invested to distill patterns from the motion of the planets known since antiquity (Mercury, Venus, Earth, Mars, Jupiter, and Saturn). His three discoveries are known today as Kepler's Laws. And they are so powerful, and so utilitarian, that barring some disaster that sets humankind back to a medieval world, Kepler's Laws will continue to be used by astronomers, navigators, and space pilots as long as our species endures.

Kepler's first law states that the orbit of a planet about the Sun is an ellipse (Figure 1.9). Kepler's second law states that the line joining a planet to the Sun sweeps out a constant area per unit of time. Sounds a little thick, perhaps, but from a pragmatic standpoint, rule number two simply means that planets move around the Sun more slowly when they are relatively far from the Sun, and more rapidly when they are closer to it. Kepler's third law describes a simple equation, a recipe, for calculating the length of a planet's "year" from the mass of the Sun and the planet's average distance from it.

In the three and a half centuries between the discovery of Kepler's Laws and their application to Pluto, a considerable body of theory and observation had accumulated concerning the inner workings of orbits and the architecture of the solar system. By the late eighteenth century, for example, it had become possible to calculate the orbit of a planet from a few well-separated measurements of its position against the stars, and to use the orbit solution so obtained to then predict with high accuracy where the planet would be on any future date. This calculation isn't hard for silicon computers, but it's no trivial affair for computers made of flesh and blood. Why? Because in addition to the force exerted by the Sun's gravity, each planet's gravity also slightly affects the motion of the others. These other planetary tugs help explain why Kepler's calculations were almost—but not quite—exact.

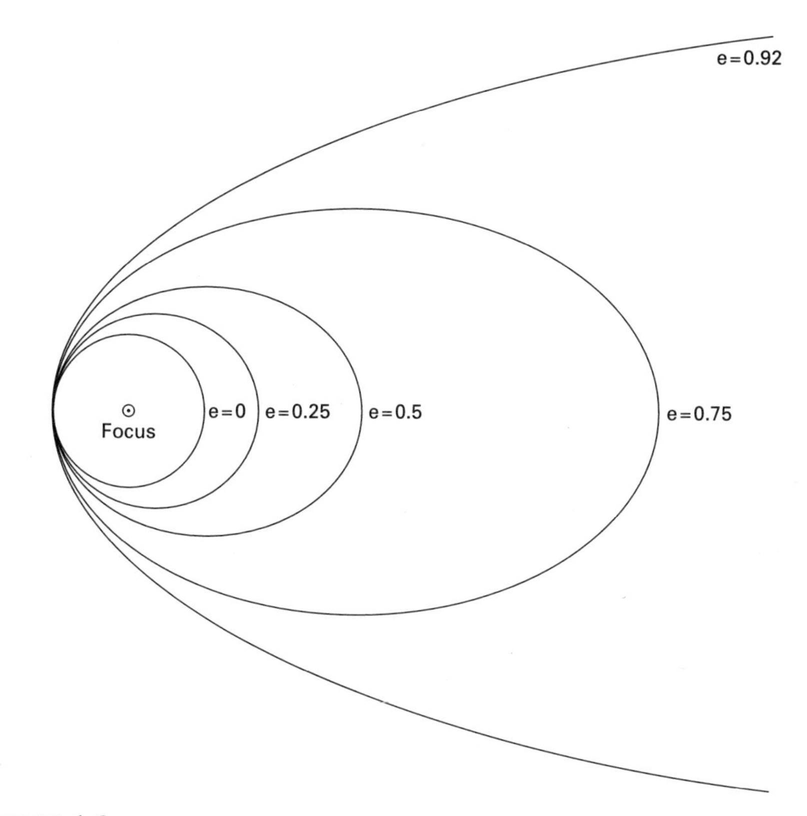

FIGURE *1.9*

As Johannes Kepler discovered, the orbits of the planets are ellipses, not perfect circles. This diagram shows a circle (zero eccentricity) and four ellipses of different eccentricity. The more elongated an ellipse, the greater its eccentricity, *e*. The point labeled "Focus" is the center of the circle and also one of the two foci of each ellipse. When a planet travels around the Sun in an elliptical orbit, the Sun is at one focus of the ellipse, which is not the center point (unless the orbit is perfectly circular). The ellipse with eccentricity 0.25 is about the same shape as Pluto's orbit. The ellipse with eccentricity 0.92 is so stretched out that only part of it can be shown here. It is similar in size and shape to the very elongated orbits followed by comets such as Halley, with periods of many decades or longer.

Against this background many of the world's astronomers set out in 1930 to learn Pluto's orbit and compute its future path. Before revealing the surprises that they found, however, it's worthwhile to take a little detour, a digression if you will, to see the forest of the solar system as a whole, before we focus on the solitary sapling called "Pluto."

EMPTY OF EMPTIES

In overview, the solar system is laid out with the Sun at its center, and the orbits of the widely spaced planets concentrically arranged outward, from Mercury to Neptune. The four inner planets move in tight orbits close to the Sun. These four, so-called terrestrial planets share broadly similar qualities of size and rockiness. The smallest, Mercury, has a diameter of 3,050 miles (4880 km); the largest, Earth, has a diameter of 7,970 miles (12,753 km). The closer a planet is to the Sun, the more rapidly it completes its orbit. Mercury, for example, lying only about 40% as far from the Sun as the Earth does, completes each circuit in just 88 days.[7] Venus, which is 70% as far out as Earth, requires 225 days to make each lap, and the Earth, of course, takes 365 days. Moving farther out, Mars takes about 1.9 Earth-years to complete an orbit.

Venus and Earth orbit on ellipses barely distinguishable from circles, but Mercury and Mars follow more exaggerated ellipses that range farther afield. Each orbit's deviation in distance from the Sun is called its "orbital eccentricity" (astronomers could also call it an orbit's "egginess," but this would lose a bit of the professional touch). The eccentricity of Earth's orbit is a little under 2%. Venus's orbital eccentricity is less than 1%, but Mars's is near 10%, and Mercury's is almost 20%.

Each of the inner planets is rocky, and each except Mercury has a substantial but razor-thin atmosphere surrounding it. Mercury's atmosphere is but a gossamer corona.

Beyond Mars, the final outpost of the four terrestrials, lies the asteroid belt, and then the vast and unabashedly bizarre wilderness inhabited by the four giant, outer planets: Jupiter, Saturn, Uranus, and Neptune. The giants are enormous (Figure 1.10), cold worlds with poisonous atmospheres a thousand times deeper than Earth's thin but refreshing shell of blue. Jupiter, the largest of the giants, could fit some 1,400 Earths within itself. Even the smallest of these Goliaths, Neptune, could itself swallow over 60 Earths.

The four giant planets travel in orbits (Figure 1.11) that are all circular to within 5%. Their orbital periods (i.e., their own years) range from 12 Earth years for Jupiter to 167 years for Neptune. Jupiter and Saturn are bright enough to easily see by eye, and they were known to the ancients as special "stars" that moved across the sky, like Mercury, Venus, and Mars. By contrast, Uranus and Neptune are much fainter and were discovered only after the invention of the telescope and its application to astronomy.

---

[7]By days we mean Earth days; each planet's own "day" has a different length.

FIGURE *1.10*

The relative sizes of the nine major planets and the Sun are shown here to scale.

The orbits of all eight planets from Mercury to Neptune lie close to a mathematical plane, an imaginary flat sheet extending out into space near the Sun's equator, which astronomers call the "invariable plane." The four inner planets all orbit within 6 degrees of the invariable plane. The four giant planets, beginning with Jupiter, orbit even closer to the invariable plane—none strays from it by more than 2½ degrees.

Now consider the accompaniments of the planets: their rings and moons. Here, there is another major difference between the inner and the outer planets. The inner planets are not encircled with rings, and

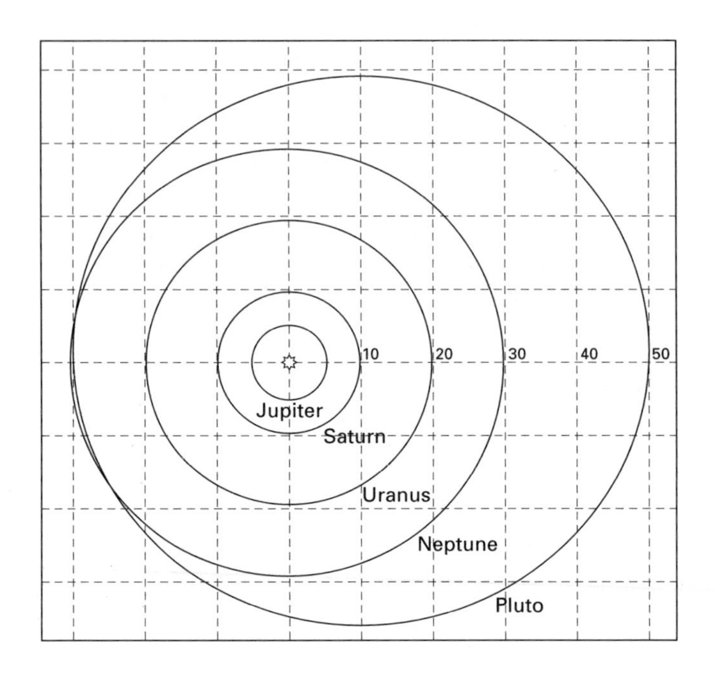

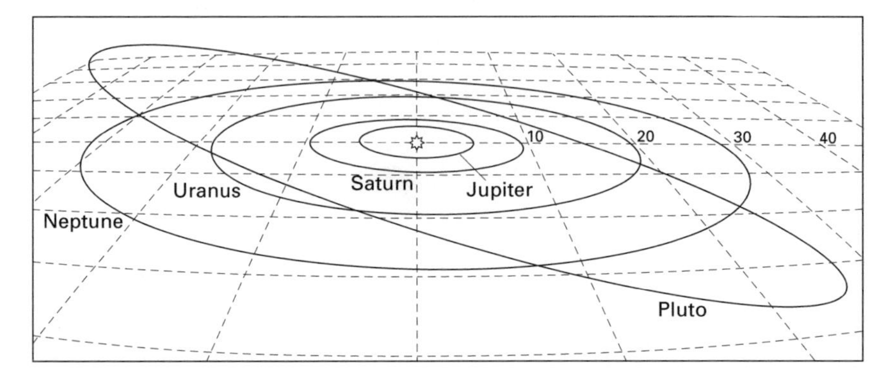

FIGURE *1.11*

The layout of planetary orbits in the outer solar system: The grids are labeled in astronomical units. Looking down on the plane of the solar system (top) shows how Pluto's strongly elliptical orbit crosses inside Neptune's. The perspective view (bottom) shows the inclination of Pluto's orbit relative to the orbits of the other outer planets. The orbits of the inner planets, including Earth, are all far closer to the Sun than Jupiter and are not shown here.

have among them just one large satellite (our Luna) and two tiny moons hardly larger than a respectable English county (Mars's Phobos and Deimos). In contrast, each of the giant planets displays a system of rings, and each possesses a dozen or more moons. Among the more than 50 moons discovered around the giant planets to date, five are larger than our own moon, and three are larger than Mercury.

When we step back from these details, these trees that form our solar system's forest, something even more fundamental emerges. The central, overarching aspect of the planetary system, which we all so casually overlook, is that it is almost completely empty. The planets themselves are but specks, spaced across millions and even billions of miles of nothingness. If the Earth were reduced in size to a simple, blue basketball, the Sun (itself then about 100 feet in diameter) would lie about 12,000 feet away, with nothing but the little orbs Venus and Mercury occupying the space between the two. In this miniature, Jupiter, about 12 feet in diameter, would circle the Sun 10 miles away, followed by Saturn 20 miles out, Uranus some 40 miles out, and lonely Neptune, 60 miles distant from the tiny yellow hearth, glowing down below. Thus, within the 80 miles corralled by Neptune's orbit in our Tinker-toy™ model, there is nothing but the nine familiar, balloon-sized planets, their Sun, and their tinier-still retinue of satellites, as well as a few-thousand scale-model asteroids (most no larger than sand grains), and perhaps a few-hundred comets, each, like the asteroids, barely a grit of sand against the yawning, empty vacuum. Empty upon Empty.

It is the magnitude of this emptiness that is so disturbing and so effective in its quarantine of worlds from one another. The solar system is, almost but not quite, nothing *but* emptiness. Which, of course, is why we call it "space."

ROGUE

Engulfed within the wild black emptiness, yet another billion miles in the distance beyond Neptune, Clyde Tombaugh had discovered Percival's planet. Our little dyevoshka, our Pluto.

Once Pluto's discovery and position became public knowledge, astronomers across the Americas, and Europe, and Asia, and Africa began to collect frequent observations of its position, so that an accurate orbit could be determined.

Two types of data came in: Most came in the form of positions from new photographs, but data also came from the astronomical libraries of the world, as various groups recognized that they had unwittingly photographed patches of sky containing Pluto over the previous 15 or so years. This kind of "archival" research is now common in astronomy, but in the 1930s, it was relatively novel.

The prediscovery observations of Pluto found in astronomical libraries were particularly useful because they greatly extended the time base over which accurate positions could be found, and they rapidly gave rise to an accurate orbit. By the early summer of 1930, some 136 archival images of Pluto had been located on old plates taken before Pluto's discovery. The most ironic of these were located by Seth Nicholson, who found that Milton Humason's old winter–spring 1919 search for Pickering's Planet O had in fact netted four images in which Pluto faintly appeared. Unfortunately, Humason had not recognized the planet, in part because he had imaged it *away from* the opposition point, where planet detection would have been easiest and least ambiguous . . . . Would that Pluto had been discovered then, and Pickering and young Humason had been the story's heroes . . . . But this was not the history that unfolded.

Armed with accurate positions stretching back as far as 1914, several astronomers calculated and published solutions for Pluto's orbit in 1930 and 1931. As a result, within about a year after Pluto had been found, its orbit had been reliably established and confirmed. This is what they found:

- Each Pluto orbit takes 248 years. When Pluto was discovered it was heading toward its closest approach to the Sun, called its "perihelion," which it later reached in 1989. Its previous two perihelia occurred in 1741, when George Washington was only a boy, and in 1493, just months after Columbus had reported his first explorations to the Reyes Católicos. Pluto follows a long orbit indeed.
- While the other planets of the outer solar system follow ellipses that are typically only 1 or 2, or at most 5% eccentric, Pluto's orbit stretched ellipticity to a new record among all the known planets—25%! As a result, on each orbit, Pluto strays over a range of heliocentric (i.e., solar-centered) distances from 3 to 5 billion miles.
- Whereas all the other planets orbit within a few degrees of the invariable plane, Pluto's orbit was tipped 16 degrees away, far askew the others.

- Pluto's path is so strange, in fact, that it actually crosses the orbit of Neptune near its perihelion, making it temporarily planet eight (not nine!) for a few decades every 2½ centuries!

Together, these findings painted a picture of a nearly fantastical orbit, unlike that of any other planet. No other planet crosses inside another's orbit. How dare planet Pluto do that, as if it were some enormous comet? How? Why? What kind of thing was this that roamed the frigid outback of the planetary system? It would take the remainder of the 20th century to find out.

CHAPTER

2

FIRST FACTS

Every book of descriptive astronomy is to-day out of date.
A. C. D. Crommelin,
President of the Royal Astronomical Society
On the announcement of the discovery of Pluto,
March 13, 1930

For almost 50 years after its discovery, across the great middle of the twentieth century, Pluto's true nature and even its most basic outlines remained the secret they had always been. The technology available to astronomy was simply too primitive to penetrate the distance between us and this faint, softly reflecting world. Through the 1930s, 1940s, 1950s, and 1960s, little was revealed. By the 1970s, however, an onslaught of more modern instrumentation had begun to reveal the face and the persona of the solar system's ninth sister.

Most of the story told in this book concerns the rapid unwrapping of the gift that Pluto has revealed itself to be. But before we tell the Tofflerian tale of the accelerated, almost madcap pace of Pluto discoveries that rolled across the research landscape of planetary astronomy in the 1980s and 1990s, it's worth returning to the yesteryears of the mid-twentieth century. It was a time when even the simplest facts about Pluto were hard to uncover . . . .

THE DARK AGES

Despite the fact that the outer planets are tens of thousands of times nearer to us than the closest stars, it's often harder to make progress in studying nearby worlds than it is in stellar and galactic astronomy.

Why is that? For one thing, there are fewer things to study, and each planet is distinctly unique. As a result, it's harder to find trends among the observations and come to general conclusions. For another thing, the number of planetary researchers and the resources available to them is smaller. But most importantly, the objects themselves don't cooperate to the same degree. Stars shine, and by virtue of their internal energy sources, as do galaxies and glowing nebulae. And the light that stars and galaxies emit reveals intimate details about their nature and inner workings. By contrast, planets reflect light from the Sun, but for practical purposes they emit no light of their own. Thus, they are more impervious to the tools of observational astronomy.

Pluto, being the farthest—and the faintest—known planet, is *the* challenge of the solar system. Why? For one comparison, Pluto's apparent diameter, as seen from Earth, is 300 times smaller than that of either Venus or Jupiter. To image Pluto's surface requires the observer to resolve something as tiny as a walnut 30 miles away. In addition, the job of studying Pluto is harder than just cracking the nut that distance imposes. At its brightest, Pluto is also 14 *million* times dimmer than is Venus at its brightest, and 1 million times dimmer than Jupiter's best; indeed, Pluto is even 1,500 times fainter than Jupiter's major satellites—Io, Europa, Ganymede, and Callisto, about which so little was known until the two *Voyager* spacecraft reached them in 1979.

So what was the point of struggling with Pluto through the long years from the 1930s to the late 1970s? It was the challenge and the attraction of exploring a truly frontier world. In those five decades of the middle of the century, when many of the leaders of modern-day astronomy were either not yet born or in the awkward stages of life, it was just barely possible to learn how long Pluto's day lasted and the color of its surface. Later, in the 1970s, it was also possible to find out that it lies on its side, as Uranus does, and to discover something important about Pluto's surface composition. It was even possible to discover that Pluto has a companion. These meager gains revealed the first sketchy details of the Pluto we know today. So too, they illustrate the budding of planetary astronomy that occurred in the middle of our century. It is a story worth telling.

SHIMMER, LITTLE PLANET

They say a picture is worth a thousand words, but Pluto's small size, great distance, and faint signal conspired against all efforts to obtain even a crude picture of it for more than 60 years. Why? Because all our telescopes suffered a common, crippling handicap: They lie at the bottom of Earth's turbulent and shimmering ocean of air.

In the early 19th century, the British physicist Lord Rayleigh discovered that the physics of optical instruments work to ensure that, all else being equal, proportionately larger instruments give proportionately better resolution of fine details. As such, for example, the 100-inch reflecting telescope at Mt. Wilson in southern California should provide about 7½ times the resolution of Slipher and Tombaugh's 13-inch Pluto finder—and in fact it does. The only problem is that the light from Pluto, having traveled straight across 5 billion kilometers (3 billion miles) or more of space to reach the top of the Earth's atmosphere, becomes defocused in that last, nasty 80 kilometers (50 miles) or so it must traverse across the thin arc of Earth's atmosphere on its way to the telescope. Astronomers often refer to this defocusing as "seeing." Seeing can vary from minute to minute, from place to place, and even along different lines of sight through the sky, depending on the amount of turbulence in the overlying atmosphere.

One might think that atmospheric seeing isn't a large effect. After all, it only limits the resolution of a telescope to between 1/7,000 and 1/1,000 of a degree, or, as astronomers say, between 0.5 and 3 arcseconds. (An *arcsecond* is 1/3600 of a degree.) It doesn't sound like much, and your eye can't usually detect it, but if you turn a telescope toward the surface of the Moon, the shimmering effect of seeing becomes immediately apparent.

For Pluto, whose tiny size and great distance conspire to give it an apparent diameter of only about 0.1 arcsecond as seen from Earth (1/36,000 of a degree), the result is disastrous. It doesn't matter whether you use the 13-inch Pluto finder at Lowell, the 100-inch at Mt. Wilson, or the 400-inch Keck telescopes atop Mauna Kea in Hawaii: The atmosphere prevents Pluto from being resolved, and that means you can't see any details on its surface—whatsoever.

Today we have begun to beat this problem by using sophisticated computers, lasers, and optical correctors to measure and then take out most of the atmosphere's twinkling, or by putting our telescopes into orbit, à la the Hubble Space Telescope. However, those options weren't practical until almost 60 years after Pluto's discovery. As a result, it wasn't possible for any telescope, regardless of the quality of its optics or its exquisite guiding system, to reveal a picture of Pluto like the ones of the closer planets, such as Mars, Venus, and Jupiter. That inability irreparably stunted studies of the ninth planet for decades.

No picture, no story—well, almost. How, for example, without resolving the disk of the planet, can one map its surface or tell if clouds might be in its atmosphere? Without features to track, how would anyone even determine the length of the planet's day? Fortunately, astronomers developed a kind of poor-man's substitute for imaging that can shed light on some of these very basic questions. The technique is called lightcurve analysis.

THE PULSE OF A PLANET

A *lightcurve* is a graph of how the brightness of a star or planet varies with time. Astronomers make a lightcurve by repeatedly measuring the brightness of an object, say every few minutes or hours, and then plotting the results. If the resulting graph is analyzed properly, it is possible to determine a number of useful things, such as the length of a planet's rotation period (i.e., its *day*), the tilt of its rotation axis, whether the planet has complex weather, and just how *contrasty* (i.e., how varied) its surface might be.

This is easier said than done because, although some stars vary by factors of 10, or even 100 on regular intervals, most planets only vary in brightness by a few percent as they rotate. As a result, the brightness measurements have to be highly precise. Additionally, all of the brightness measurements must either be made under the same circumstances of weather and atmospheric turbulence, or carefully calibrated in comparison to an unchanging star (having a constant brightness). Such a comparison works because the atmosphere affects both astronomical objects (the target object and the unchanging star) in the same way. Thus, if an astronomer chooses a star that is known to have a constant brightness, the star can be used to compensate for the vagaries introduced by Earth's atmosphere, which astronomers must tolerate whenever they observe. More subtle complications also arise, but the general idea is as we just described it.

How exactly is the brightness of the planet and the comparison star to be measured? One well-established technique for obtaining lightcurves developed around photography, which became a standard data-recording medium for astronomy as early as the 1850s and 1860s. One advantage of photography over visual records is that it's less subjective. A second advantage is that a photograph can be analyzed by different astronomers to cross-check the results. The most important advantage, however, is that a photograph can measure the brightness of fainter objects than the eye can see. How? By leaving the shutter open for a period of time—by exposing the image "deeply"—it becomes possible to record faint details.

Without any doubt, photography brought about a scientific revolution in astronomical research during the century between 1850 and 1950. It allowed the science of astronomical observing to move from sketches and drawings made by humans describing qualitatively what they saw at the back end of a telescope, to a reliable, repeatable, easily reproducible product that could be *quantitatively* analyzed to reveal subtleties no sketch could show. Among its many successes, photography made possible the mapping of faint nebulae, the recording of spectra for chemical analysis, and the discovery of myriad asteroids—and distant, slowly moving Pluto.

By the mid-20th century, there were many specialized photographic emulsions for astronomical use, but still all astronomical photography suffered a common weakness, called "nonlinearity." What a great word—nonlinearity. Toss it out at a party sometime.

What astronomers mean by this 10-penny piece of jargon is actually straightforward: If two objects x times different in brightness are imaged onto the same photograph, their resulting images will likely not be precisely x times different in brightness. They may be close, but they won't be *exactly* proportional. As a result, accurate, quantitative comparisons become difficult.

Nonlinearity also rears its putrid head another way, when the same object is exposed at different times on different emulsions, or even with the same kind of emulsion but with different exposure times or different camera temperatures. Nonlinearity is a plague that infects lightcurve (and other) quantitative studies that use photography because the uncertainty in the measure of a planet's brightness on a photograph can be as large or larger than the lightcurve of the planet itself. These and other difficulties of working with photographs prevented *anyone* from obtaining a useful lightcurve of Pluto in the 1930s and 1940s.

Although photography revolutionized astronomical capabilities a century ago, very few astronomers today ever actually deal with photographs. Photography has been almost entirely replaced by a far more accurate technique using compact solid-state CCD (charged-coupled device) cameras, which will be described in Chapter 3. Unfortunately, when Pluto was discovered in 1930, the solid-state devices we take for granted today didn't exist. However, by 1950, the electronics revolution was beginning, and an early device called a photomultiplier tube (or PMT) was becoming widely available. Because of their advantages over photography, PMTs were popular for making accurate light level measurements, which astronomers call "photometry."

PMTs aren't like cameras. They are just simple light meters that don't reveal an image of their subject—but only a measure of its brightness. The good news about PMTs is that they produce far more linear results than photographic emulsions do. An x times brighter object will give an x times brighter signal. What an improvement! What a relief! Furthermore, because they amplify the signal they detect manifold, PMTs are much more sensitive than photographs. The bad news is that PMTs are large and bulky, as shown in Figure 2.1. They are so bulky in fact that usually just one PMT fits at the place where the camera goes at the back end of the telescope. As a result, instead of getting a two-dimensional image with the thousands of resolution cells that a photograph gives, a PMT gives you just one *pixel*—one picture element. There is no image, just a single measure of the brightness of the thing you're looking at. Point it at Jupiter, and you get the total brightness of Jupiter. Put it at the back end of a telescope pointed at the Moon and you get the brightness of the little plot of land on the Moon the telescope sees, but not an image of that spot, just its total brightness.

Fortunately for lightcurve work, Pluto is too far away to be resolved into a disk when viewed with ground-based telescopes, so it doesn't much matter that the PMT doesn't give an image. All we need to form a lightcurve is an *accurate* record of how the brightness of the whole of Pluto changes over time, and that's what a PMT can provide.

So when PMTs began being applied to astronomy in the early 1950s, Bob Hardie of Vanderbilt University began a project to measure Pluto's lightcurve. If he succeeded, Hardie knew he would obtain the first physical information on Pluto since its discovery. Two false starts on Hardie's part resulted only in evidence for Pluto's variability—but even this was an encouraging sign because many planets (including Mercury, Venus, Uranus, and Neptune) didn't show detectable variations in brightness . . . just completely flat lightcurves.

Hardie was joined by a colleague, Merle Walker, and together they set out to improve on Hardie's initial results. They first observed Pluto together in early 1954, and then, in the spring of 1955, the pair secured observing time on the largest telescope at Lowell, a 42-inch reflector. As they collected their data, Walker and Hardie reduced it to a common standard, using two nearby stars as their reference post. Within a few weeks, it became clear that Pluto was varying much more than most planets—indeed, its brightness fluctuated by more than

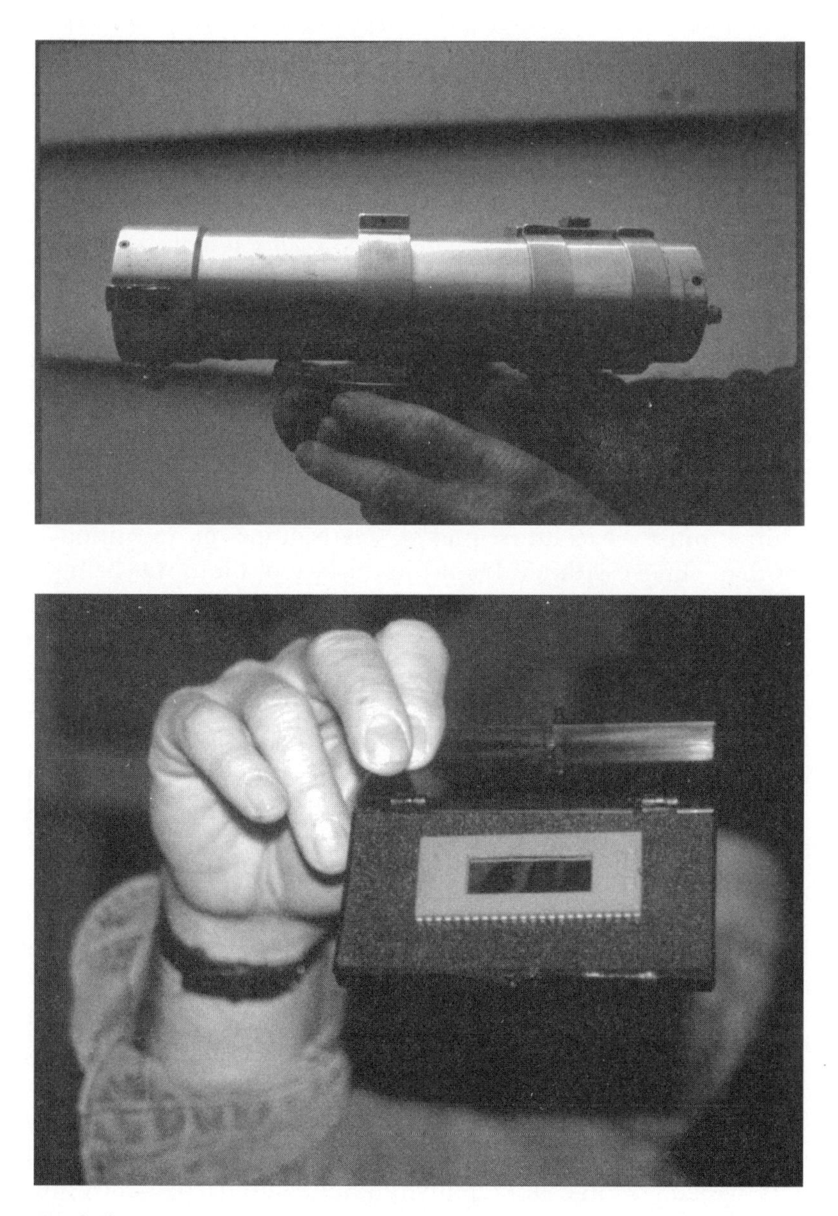

FIGURE *2.1*

A photomultiplier tube (PMT) such as the one shown on the top, in its metal housing, is capable of recording just one item (or pixel) of data—the total intensity of all the light falling on it. The main advantage of a PMT over the analysis of the brightness of an object on a photograph is the PMT's greater accuracy for most circumstances. By comparison, the smaller charge-coupled device (CCD) shown on the bottom panel records the intensity of light on each of the many thousands or even millions of individual tiny pixels that make it up. In this way, a CCD can produce a complete two-dimensional digital image, similar to a photograph, but it records the variations in the intensity of light across the image much more accurately than a photograph does.

10%. Even more surprisingly, the brightness fluctuations repeated themselves like clockwork, with a period of 6.39 Earth days, plus or minus a meager 4½ minutes. A decade later Bob Hardie wrote a popular account of these data in *Sky & Telescope*:

> . . . *During each rotation, the planet's light increases for about four days, then drops to a minimum again in about two. If Pluto is assumed to have a more or less round disk, evenly illuminated out to its edge, the lightcurve asymmetry can be explained only by an extremely unlikely pattern of dark and light surface markings.*

The information Hardie and Walker obtained, though meager in quantity, was stunningly revealing. Clearly, Pluto's variability was remarkably large, larger in fact than for any of the other planets except Mars. As the *Sky & Telescope* article revealed, even Bob Hardie found it hard to believe how stark the surface markings implied by the lightcurve must be. Pluto's precise metronome of repetition also meant that, almost without doubt, the surface of Pluto was being seen through little if any atmosphere. Certainly if there were an atmosphere, then it didn't include highly variable clouds, for if it did, the lightcurve would have displayed an ever-changing set of bumps and wiggles on it. Further, the shape of Pluto's lightcurve was studded with repeatable little hills and dales that hinted at a potentially complex set of surface markings just waiting to be revealed, if only Pluto could be imaged. Finally, the lightcurve's long period, 6 Earth days, and 9.4 hours, was itself remarkable—no planet beyond the orbit of the Earth except Pluto takes more than 25 hours to rotate. Indeed, the gas giants that dominate the outer solar system each spin round with whirlwind days ranging from 10 to 20 hours. Why, astronomers wondered, did Pluto rotate so slowly?

These facts carried with them the impact of an artillery shell landing among the planetary scientists of the day. Here, on the ragged outer edge of the planetary system, was a smallish world, rotating at glacial pace, and possessing either little atmosphere at all or, at least, a nearly cloud-free one. These are not like the attributes of the other outer planets, which are large and rapidly rotating, with deep, cloudy atmospheres. Pluto was revealing itself to be a very different kind of beast.

A COCKEYED WORLD

The discovery of Pluto's strong, misshapen lightcurve provided an important attraction to planetary astronomers of the 1960s and 1970s, and spurred forward efforts to apply newer and more difficult types of measurements to faint Pluto. It also encouraged researchers to improve the lightcurve data in an effort to yield further results.

New lightcurves were obtained in 1964 by Hardie himself, and then in 1972 by Leif Andersson and John Fix. From these, it was discovered that Pluto's lightcurve was changing in two ways. First, the average brightness of the surface was gradually falling, which indicated that Pluto was dimming, even though it was moving closer to the Sun! Second, the *amplitude* of Pluto's lightcurve—that is, how much it varies from its peak to its minimum—was increasing with time. The decrease in Pluto's overall reflectivity was a small effect, but as shown in Figure 2.2, the increase in lightcurve amplitude was not: It doubled between 1954 and 1972.

Bob Hardie conjectured that Pluto's surface might be changing its reflectivity due to the slow movement of surface frosts as the planet approached perihelion. After all, Pluto had moved almost a billion miles sunward since its discovery. But wasn't this exactly the opposite effect

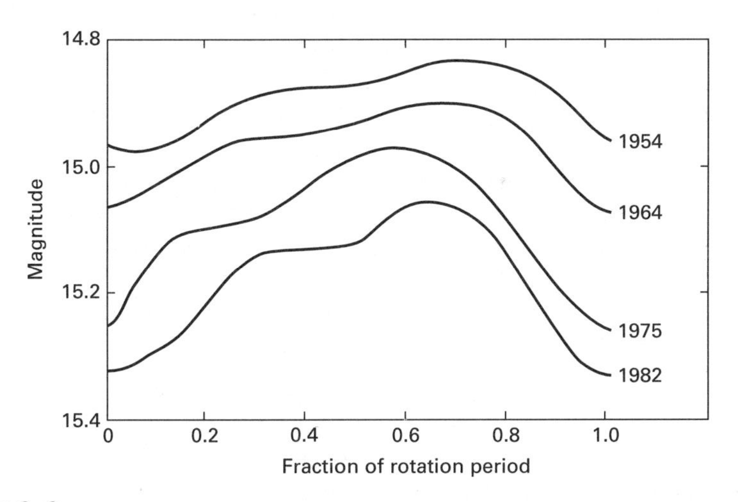

FIGURE 2.2

The change in Pluto's lightcurve over almost 30 years, from 1954 to 1982: During this time, Pluto became dimmer, and its variation in brightness over its 6.4-day rotation cycle became more extreme. (Brightness is shown here on the magnitude scale, in which larger numbers mean a fainter signal.) To ensure a fair comparison, these four lightcurves have been adjusted to correspond to Pluto being illuminated by the Sun in exactly the same way and at the same distance from the Sun. This adjustment proves that the observed change over time is the consequence of a real change in the brightness of the visible surface of Pluto. (Adapted from Stern et al., *Icarus, 75,* 1988)

that common sense might suggest? Shouldn't Pluto have grown brighter as it moved closer to the Sun?

Andersson and Fix proposed an alternative interpretation, which explained both the change in Pluto's average surface brightness and its increase in lightcurve amplitude. Because the lightcurve seemed to maintain its gross shape even as the difference between the minimum and maximum brightness increased, they reasoned that the tilt of Pluto's axis was probably large. If that were in fact so, then it would cause different bands of latitude to be visible at different times as Pluto moved around the Sun and was seen from different perspectives. The tilt of a planet's spin axis, relative to its orbital plane, is called its "obliquity." Earth's tilt, or obliquity, is 23.5 degrees. It is just this tilt that generates our seasons. Among the planets from Mercury to Neptune, obliquities range from Mercury's nearly 0 degrees, to Uranus' 98 degrees; in fact, except for Uranus, which is literally lying on its side as it orbits around the Sun, all of the inner eight planets rotate with 30 degrees or less tilt. Andersson and Fix observed that, like Uranus, Pluto's obliquity seemed to be very large.

The two astronomers constructed a computer model to see what values of the obliquity could explain the trend of varying lightcurve shape. What they discovered was that Pluto's obliquity had to be large—at least 50 degrees, and perhaps quite a bit higher still. Further, they concluded that the angle from which we are viewing Pluto, called its "aspect angle," had moved equatorward from 1954 to 1972. The decreasing brightness of the planet over this 18-year time span indicated that Pluto's polar regions are brighter than its equatorial regions.

Did this mean that there are polar caps on Pluto? Perhaps—but the case was pretty meager and the evidence very indirect. It was also possible that Pluto's appearance had changed for other reasons, such as some kind of actual change in the patterns of markings on the surface. To determine whether there were in fact polar caps would require more and better evidence—after all, at the time it wasn't even clear whether there were ices *anywhere* on Pluto's surface. To search for ices it was necessary to explore Pluto's surface composition, which calls for a technique known as "spectroscopy."

READING THE RAINBOW

Thanks to color vision, all of us carry through life a crude but essential spectrometer within each eye. Thanks to this, our eyes recognize that light comes in a range of wavelengths, displayed in a delightful panoply of reds, yellows, blues, and all the combinations in between. Color perception, like our other senses, continuously conveys important information to us, and, at the same time, adds a wonderful richness and aesthetic quality to our lives. Given the experience of color vision so deeply rooted within us, it is not surprising that astronomers invented a way to precisely measure the color of the distant stars and planets as a means to understand them.

The basic technique invented to measure colors is easy: Simply make two images of the object of your interest using filters that pass different colors—for example, red and blue. If the object is brighter in the red-filtered image, then we say its color is red because it must be emitting or reflecting more red light than blue light. Similarly, if the bluer filter gives the brighter image, then we say the object's blue. If the two filters yield equally bright images, however, then the object doesn't have much intrinsic color, and astronomers call it "gray."

Very nice, very straightforward—in principle. However, color measurements have some subtle idiosyncrasies that must be carefully taken into account. For example, different color filters transmit light with different degrees of efficiency. Also, the detector used to record the light usually has better sensitivity at one end of the spectrum (either red or blue) than at the other, depending on its type. If, for example, the detector is most sensitive to blue light, then it might register a larger signal through a blue filter than a red filter, even though the object's true blue brightness is weaker than the object's real red brightness. These and other factors make the game a bit more complicated, but then, part of the fun in this sport is knowing all the tricks and executing them flawlessly. For example, there are tables of numbers prepared for each filter to correct for each one's efficiencies; it is also possible to calibrate detectors, or even the whole observing system, by looking at either a test pattern or a distant object with well-established colors.

In the very month that Pluto was discovered back in 1930, Carl Lampland had compared the ratio of Pluto's brightness through yellow and blue filters against the ratio he measured using the same filters for Neptune. This type of comparative, or relative, measurement is self-calibrating, in that all of the hard parts of the calibration turn out to be the same for Pluto and Neptune. If there is a difference between Pluto and Neptune, this simple trick shows it immediately. Using a relative measurement like this, there is no need for complex and careful corrections for instrumental factors. By comparing Pluto with Neptune, Lampland found that Pluto was much yellower than its bluish sister.

He attributed the different hue to "a . . . [kind of] different atmosphere." Lampland didn't know how right he was.

Knowing that Pluto is yellower than Neptune is a nice comparative, but it doesn't say what Pluto's true color actually is. Is it simply that Pluto appears yellowish because it is mirroring sunlight? To answer that question, one additional factor had to be taken into account: Pluto's apparent color (like that of every other planet in the solar system) is influenced heavily by the color of the yellow Sun, which is the light source illuminating it. To determine Pluto's true color, it would be necessary to measure Pluto's color and the Sun's through the same observational setup.

How does this work? Consider these examples in an imagined experiment using a pair of red and blue filters: Suppose that the Sun is twice as bright in the red-filter image as in the blue-filter image, but Pluto is four times as bright in the same red-filter image as the blue-filter image. In this situation, Pluto is clearly redder than the Sun. Now, suppose that Pluto—like the Sun—is only twice as bright in the red-filter image as in the blue-filter image. If this were the case, then Pluto would be reflecting the sunlight without changing its color, so astronomers would say that Pluto reflects all colors equally and is thus gray. As a third alternative, suppose that Pluto is only a little brighter through the red filter than through the blue filter, while the Sun is twice as bright in the red filter as through the blue one; in this scenario, Pluto itself must be rather bluish, so that Pluto can tone down the red color of the Sun when it reflects sunlight back to Earth.

In 1933, Vesto Slipher reported on exactly this kind of measurement. Less than 3 years after Lampland's comparison with Neptune, Slipher had something more meaningful: "Pluto's light in the yellow and red is apparently somewhat stronger than normal sunlight, and so he (*sic*) is a yellowish planet, possibly standing between Mercury and Mars in color."

Over the years, astronomers recognized that color measurements could be intercompared much more easily if everyone used the same kinds of filters. So, a standard set of five filters was developed to measure the brightness of stars and planets in wavelengths ranging across the spectrum from just beyond the bluest wavelengths the eye can detect to a little bit beyond the reddest. These five filters are known as U (ultraviolet), B (blue), V (visual—actually, green), R (red), and I (near-infrared). Comparing the brightness of an astronomical object through any two of these gives a quantitative measurement of its color, known as "a color index." By the early 1960s, good UBVRI measurements of

Pluto and many other objects in the solar system had been made by people such as Bob Hardie, and it was possible to say rather precise things, such as "Pluto's B-V index is near 0.85." Sounds very official, (and it is) but it just means that Pluto is just a skosh redder than the Sun, and it would appear to the eye only faintly more orange than does a full Moon—which means, hardly at all!

Now that we've seen how color measurements are made, it's worth asking what colors tell us about Pluto's fundamental nature, or about its surface composition. As it turns out, not much. Colors give a feel for how a planet may appear, but they just don't provide enough specifics to deduce hard information on the composition of the object's atmosphere or surface. After all, objects as different as our sky and a blueberry are blue. And despite their vastly different nature and composition, both bananas and the Sun are yellow. Likewise, Mars is red as are roses, bears, and many kinds of beers, but these objects don't bear much compositional similarity to one another. Mama always said you can't judge a book by its cover.

To learn something definitive about compositions, much more detailed information than just color is needed. What is required is a *spectrum,* which is simply a detailed graph of the brightness of an object across a range of finely divided colors, or wavelengths. The idea is this: Measure the brightness of an object at 100 or 1,000 colors between red and blue (or better yet, between infrared and ultraviolet) wavelengths, and you'll have something you can sink your teeth into. Why does this work? Because every type of atom and substance imprints its own spectral signature on the light that it emits or reflects. These signatures come in the form of little dips called "absorptions," which occur at various specific wavelengths (i.e., colors) unique to each molecule. In effect, the pattern of absorptions for each molecule forms its fingerprint. If a given molecule is sufficiently prevalent on the surface or in the atmosphere of a planet, it can make deep enough absorptions to be seen in a spectrum.

Soooo, if you want to know what kinds of material lie on Pluto (or Titan, or Triton, or Timbuktu for that matter), just compare its spectrum to those of simple materials, and see how much of each material is required to obtain a good match (see Figure 2.3). It's something akin to reverse-engineering a kitchen recipe: "We'll need a lot of salt, a little paprika, and a little pepper, but hold the sugar, cinnamon, and sage." The technique is called spectral matching, and it is so powerful that whole libraries of templates for pure materials, called "reference spectra," have been measured and then carefully catalogued on computers to make the job easier.

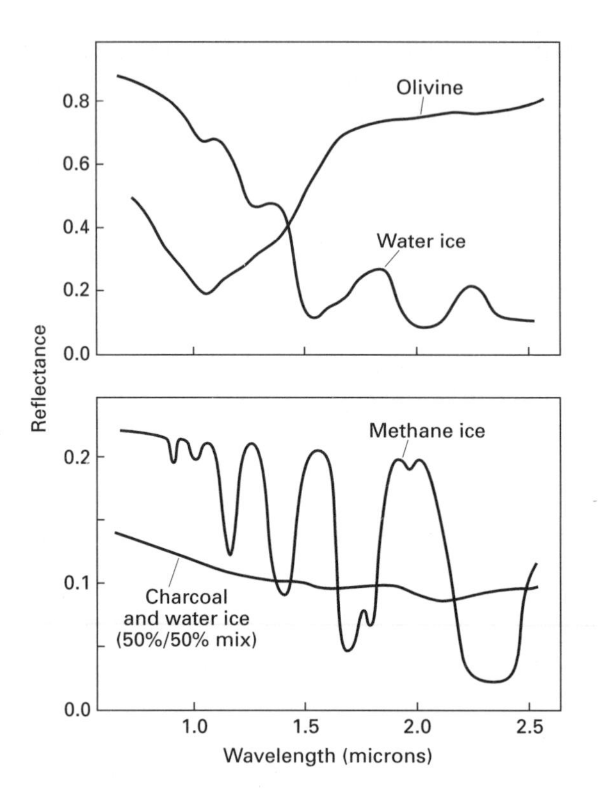

FIGURE *2.3*

The characteristic spectra of infrared radiation reflected by different materials is an important diagnostic tool for identifying the surface composition of planets and their satellites. These examples show the very different reflectance spectra of olivine (a common silicate material), water ice, methane ice, and a mixture of water ice and charcoal (50:50 by weight), and they demonstrate the basics of how the infrared spectra of common materials can be used to identify surface compositions across the solar system. (Note the expanded scale of the bottom panel.)

Unfortunately, the technology for making spectroscopic measurements of objects as faint as Pluto didn't come along until the mid-1970s. Why? One reason is that, in order to record a spectrum, the light from the target body must be divided up among many closely spaced colors (usually called "channels," or "bins"). Therefore it takes a lot longer to obtain a spectrum than to obtain a simple color image through a broad filter that lets much more light through. Moreover, to avoid the nonlinearity problems associated with photography, the only option in the 1960s and 1970s was to use a PMT detector. However, as we mentioned above, a PTM has only one pixel. This means one can only use it to measure the light from one spectral color (or bin) at a time. As a result, it is necessary to measure each wavelength in turn until the entire spectral region is covered. Therefore most astronomical spectrometers of the 1950s through 1970s were very slow at measuring an entire spectrum. It could take all night—or several nights—to build up a strong signal.

The first published spectrum of Pluto was reported in 1970 by a small group of University of Iowa researchers led by John Fix, who used their 24-inch telescope at Hills, Iowa. Fix and company's telescope was too small to allow them to get a good result if they broke up the light into too many narrow slices of the spectrum, so all they could manage was a 21-point spectrum from blue to red wavelengths. Limited as that seems, it was a step forward.

The spectrum that Fix and friends obtained is shown in Figure 2.4. Notice how this spectrum shows that Pluto is brighter at the red end of its spectrum than at the blue end. This is due to Pluto's reddish color, as had been that previously discovered by Hardie's UBVRI photometry.

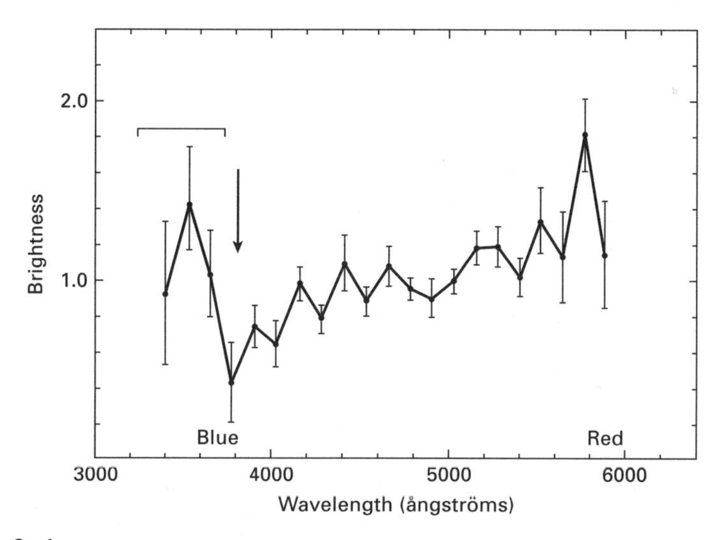

FIGURE *2.4*

The first spectrum of Pluto, obtained in 1970 by John Fix and his colleagues. They measured the brightness of Pluto at 21 points across the visible spectrum. The vertical bars through the points are "error bars," indicating the extent to which each measured point is uncertain. The general trend shows that Pluto reflects more red light than blue light. The three points on the far left (bracketed) turned out to be mistakenly high, making it appear as if there is an absorption dip in Pluto's spectrum (arrowed). (Adapted from Fix et al., *Astronomical Journal, 75,* 1970)

The second thing to notice about Fix's spectrum is that the three leftmost (i.e., bluest) channels bump up, compared with their redder neighbors. One zealous astronomer interpreted the sawtoothlike shape of the dip between the red slope on the right and the three leftmost channels as an absorption feature due to either iron-bearing minerals, or possibly silicate rocks, and happily went off to write a scientific paper claiming that Pluto's surface had copious iron or silicates on it. Nice try, but he was (how do we say it?), well, . . . overinterpreting the data. Simply put, the dip attributed to an absorption feature wasn't real at all—it was caused in large part by the fact that the spectrometer Fix used wasn't as sensitive to very blue light and very red light as it was to light of intermediate colors. As a result, the error estimates (those little verticle bars on each measurement point) are larger at the edge of the spectrum than in the middle. In fact, the apparent increase in Pluto's reflectivity at the three bluest data points isn't real either—it was just an artifact of this particular instrument, which tended to show just such an upturn at the bluest wavelengths it could record when measuring faint signals. If you want to get it right, you have to be careful, and that means knowing the detailed ins and outs of how the measuring tools you are using really work. The fellow who leapt to quick conclusions based on meager and noisy data, it seems, was a little too quick for his own good.

Better Pluto spectra in the visible region, with both greater precision and more densely spaced spectral measurements, became available as the 1970s progressed. However, the real prize waited in the infrared region of the spectrum, which has wavelengths longer than the reddest light our eyes can detect, and which is far richer in diagnostic spectral features of solid materials than is the visible band we can see with our eyes. As a result, most compositional studies of planetary surfaces now employ infrared instruments.

By the early 1970s, infrared detectors were advancing rapidly in sensitivity, largely as a spinoff from Cold War military investments in detectors for heat-seeking missiles. As the infrared devices improved, astronomers applied them to successively fainter targets: first to the large satellites of Jupiter, and then to the main satellites of Saturn. These devices revealed that most of these little worlds were encased in shells of water ice. There were also a few oddballs, like Jupiter's Io, which was practically barren of water and seemed to be littered in sulfur compounds. Equally notable was Saturn's satellite Titan, which is about the size of Mercury, and which displayed evidence in its near-infrared spectrum for a methane atmosphere.

PLUTO'S SIGNATURE

With the frontier rolling back in leaps and bounds as infrared-detector technology kept improving, the desire to obtain an infrared spectrum of Pluto was keen. And at the rate that detector-technology improvements were becoming available through military-development programs, it looked as if the feat might just be achievable by the end of the 1970s. More than one 1970s planetary astronomer put Pluto on his planning calendar for 1978.

However, those astronomers waiting for the better infrared detectors that would be in use at the end of the 1970s never had a chance. They were scooped by three young astronomers with a clever idea.

These three men—Dale Cruikshank, Carl Pilcher, and David Morrison—realized there was another way to make some inferences about the surface composition using a large telescope feeding an infrared-sensitive PMT with specially selected wideband infrared color filters in front of it. Although the information content was crude, Cruikshank and crew knew that with wideband filters, they could get enough signal through to obtain useful results right away, without waiting for a sensitive enough infrared spectrograph.

The basis of their technique depended on choosing filters in a very clever way to yield diagnostic results. Dale and his colleagues recognized that the most common frosts expected in the outer solar system—namely, water ice, methane ice, and ammonia ice—reveal themselves by strong absorption features between 1.4 and 1.9 microns (a *micron,* equal to 10,000 angstroms or 1/10,000 of a centimeter, is the standard unit of wavelength in infrared astronomy). Importantly, each of these ices has its own unique set of absorptions, as shown in Figures 2.5 and 2.6.

Cruikshank and company selected five filters that, as a set, could distinguish among the most likely ices expected on Pluto. Three, called J, K, and L, were standard astronomical filters which extend the UBVRI system into the infrared part of the spectrum. They also constructed two customized filters, which they called "H1" and "H2." The pattern of brightness measured through this set of five filters could distinguish among water ice, methane ice, ammonia ice, and common rocks.

After requesting and receiving observing time on the large, 4-meter (158-inch) Mayall telescope at Kitt Peak, Arizona, Cruikshank, Pilcher, and Morrison undertook four nights of intensive Pluto studies in March of 1976. On each night, they repeatedly measured Pluto's faint reflectivity through their five filters. After carefully calibrating the data

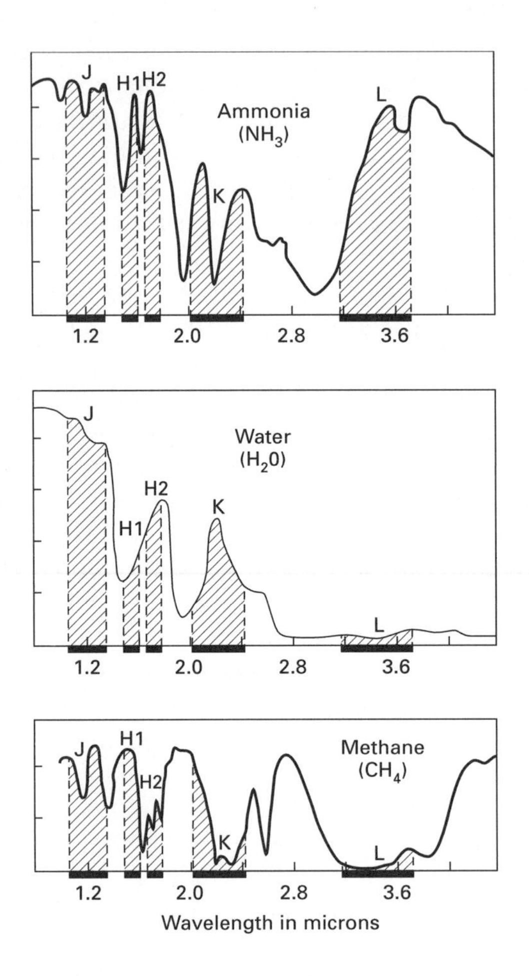

FIGURE 2.5

The first evidence for methane frost on Pluto came from measuring its brightness in infrared light through a set of filters in 1976. These diagrams show the windows corresponding to five of the six filters used by Dale Cruikshank and his two colleagues, superimposed on the spectra of infrared radiation reflected from three chemically different ices that might have been expected: ammonia, water, and methane. The relative intensity expected through each of the filters follows a different pattern for each ice. By determining which ice fitted their measurements best, Dale Cruikshank and his colleagues deduced that methane frost was present on Pluto. (Adapted from Cruikshank et al., *Science, 194,* 1976)

FIGURE *2.6*

Dale Cruikshank, who with colleagues Carl Pilcher and David Morrison, found the first evidence for methane frost on Pluto's surface. (J. Mitton)

against standard stars selected for their constant and well-known infrared brightness, they found that Pluto's surface showed no evidence for rock, water ice, or ammonia ice, but strong evidence for *methane ice*!

Cruikshank and company's simple and elegant experiment beat the first spectroscopic detection of methane on Pluto by more than 2 years, and was the scientific equivalent of the mouse that roared. For the first time, there was *definitive* information about the composition of Pluto's surface—and the three young scientists milked the data for all they were worth.

The fact that an exotic ice such as methane covers Pluto's surface, rather than water ice, or ammonia, or rock, indicated immediately to Cruikshank that Pluto's surface chemistry was far different from almost all of the distant outer-solar-system satellites that had been spectroscopically explored. Further, the fact that the methane signal in the data didn't quite match the pure methane signature the three astronomers obtained in a laboratory experiment, using the very same filters, indicated that something else was probably mixed with the methane on the surface. Further still, Cruikshank, Pilcher, and Morrison pointed out that the existence of methane, which cannot condense to an ice on planetary surfaces except in the cold outer solar system, argued against speculation that Pluto was an errant inner planet or asteroid that somehow found its way to the great deep beyond Neptune. This basket of fruitful findings echoed around the worldwide planetary community and caused a new wave of researchers to sit up and take notice of Pluto. The little planet was developing a reputation for scientific intrigue far beyond its diminutive size!

As 1976 turned into 1977, and then 1978, the long-awaited, new generation of infrared spectrometers became available, allowing far more detailed reconnaissance of Pluto's infrared spectrum. In short order, no less than three teams confirmed Cruikshank & Company's methane signature and opened the door to more detailed analysis of it. At the same time, however, another breakthrough was being made—one that no one had expected.

PRECIPICE

By the time the late 1970s arrived, the first really useful fragments of the Pluto puzzle were falling into place, and they indicated that Pluto is not at all as gargantuan as the other outer planets are.

The early evidence that Pluto is so much smaller than the other outer planets came from several quarters. First, as the suspected differences between Neptune's orbit and its measured positions evaporated, smaller and smaller estimates of Pluto's mass were required to explain them. By the time of Lowell's death, estimates for the mass of the perturber (Planet X) had dropped from perhaps 50 times the Earth's mass to just 5 or 10 times its mass. Then, when Pluto was discovered, no disk could be resolved. Although Earl Slipher's quaint demonstration showed that a small-enough planet would show no disk from a great distance, it was worrisome to many that over the succeeding 45 years, no one had ever, even during the rare moments of exceptional atmospheric clarity, seen Pluto as anything but a pinpoint on the sky—and this wasn't for a lack of trying. Astronomical giants such as Gerard Kuiper and Milt Humason tried, again and again. Kuiper even had a special machine called a "disk-meter" built to project tiny little calibration disks next to Pluto's image in the telescopes for comparative purposes. Using the disk-meter on the 200-inch Hale telescope, the largest astronomical telescope the world could then offer, Kuiper concluded that Pluto's diameter was near 5,900 kilometers (3,700 miles). This meant that Pluto was indeed smaller than the other giant planets, and indeed, smaller even than Earth, but as better measurements later proved, Kuiper's 5,900-kilometers diameter was a gross overestimate.

Other tricks to get at Pluto's true size were tried, as well. For example, could Pluto's effect on Halley's comet be measured? If so, its mass and therefore its size could be constrained. Might Pluto pass over a star? If so, then its size could be directly measured by timing the tiny eclipse. Wonderful, but no such event was ever observed until the end of the 1980s. Such ideas deserve an "A" for ingenuity but yielded little in the way of definitive results.

Through decade after long decade, estimates of Pluto's size kept shrinking. More and more exacting results gave only increasingly tight upper limits to Pluto's diameter. It was frustrating, like waiting by a phone that never rings, and which, by its continued silence, reminds one that the caller is increasingly absent.

Cruikshank's team also added to the fray by noting that laboratory samples of methane frost appear to the eye to be as reflective as fresh water-snow, indicating that Pluto's reflectivity must be very high—so high, in fact, that it seemed likely that Pluto would be still smaller than then expected. Based on the high reflectivity of methane, Cruikshank, Pilcher, and Morrison estimated Pluto's diameter was between perhaps 1,600 and 5,300 kilometers (1,000 to 3,300 miles)—smaller than the Earth's moon!

It got so bad that space physicists Alex Dessler and Chris Russell parodied the situation by plotting the mass estimates for Pluto as a function of time. Their conclusion: Pluto was disappearing, and by the early '80s, it would be gone altogether. (See Figure 2.7.) This lack of respect, albeit tongue-in-cheek, was about to evaporate . . . .

COMPANION

June 22, 1978, was one of those hot summer Thursdays in Washington, D.C., and U.S. Naval Observatory (USNO) astronomer James Christy was preparing to move his family from their apartment to a new home within a few days. Looking for a "short and routine" task to round out his week, Christy had no idea that he was about to make one of the most significant marks on the history of planetary astronomy in the 1970s.

Christy's boss, Bob Harrington, suggested that Christy "measure" a set of Pluto plates as data to feed into a better determination of Pluto's exact orbit. Because he worked for the Naval Observatory, which among its many navigation-related duties is responsible for precise planet tracking, Christy had done this kind of work many times in the past. Indeed, the plates Harrington gave Christy to measure ("measur-

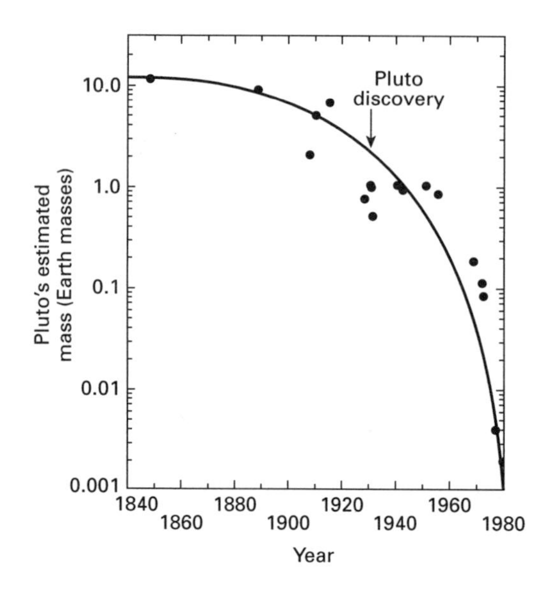

FIGURE 2.7

Before Pluto was found, and over the first 5 decades following its discovery, estimates of its mass became lower and lower. In 1980, Alex Dessler and Chris Russell jokingly insisted that a continuation of this trend would lead to the planet's imminent disappearance! It was subsequently determined, as a direct result of the discovery of Charon, that Pluto has only 2 thousandths the mass of Earth and is roughly 2,350 kilometers across. (Adapted from A. Dessler and C. Russell, *Eos, 61,* 1980)

ing plates" is a shorthand astronomers use when they mean to determine the precise position of a planet on some images), had been obtained in Flagstaff at Christy's own request that April and May, in support of a Naval Observatory program of careful planetary-orbit determinations.

In toto, there were 18 images of Pluto to be measured, but Harrington remarked it might not take long because many of the plates had been marked "defective" by the astronomers and technicians at USNO's Flagstaff station, where they had been obtained and developed.

As Christy inspected the plates using a microscope, he noticed that the images of Pluto indeed appeared strange—they were somehow elongated, as if the telescope hadn't quite guided correctly during the exposures, or the atmospheric turbulence had been particularly poor. Perhaps the plates themselves were defective; sometimes this happens . . . .

Then, slowly, Jim Christy realized something. At first it was subliminal, but then it crystallized as a conscious thought: Pluto looked elongated, but the stars on the plates looked perfectly round! The plates weren't defective—Pluto was.

On the plates taken April 13th, the extension of Pluto's image was to the south; on the May 12th plates, the extension was to the north. Maybe, Christy thought, he had discovered a giant mountain on Pluto, but the height needed to create the extension would be ridiculous: "No way."

On second thought, maybe a giant Krakatoa was erupting on Pluto? Now that would be something—no one had ever seen a volcano going off on a planet. Nah, volcanic eruptions like that don't usually last a month. Month. Month? Moonth! "What," Christy later wrote, recalling his thoughts of the moment, "Pluto has a moon?"

Could this be? Was this real? Like Tombaugh 48 years before him, Christy felt the need to have another pair of eyes see the thing (see Figure 2.8). So within minutes, he had corralled astronomer Jerry Josties into the darkened room. Josties agreed something interesting was definitely on these "defective" plates.

Christy and Josties tried to eliminate the possibility of a moon by brainstorming what else could cause a distant planet to look like this. After all, Pluto's image was less than a millimeter across—and the elongation was less than half that. It wouldn't take much to create the tiny extension Christy had detected.

Ah—some dirt on the plate at just the wrong spot! But dirt spots were very rare; and how could specks of dirt be on the Pluto images in all the plates, but not on any of the stellar images?

Not a plate flaw, . . . hmmm, well maybe it was a background star just off Pluto's side! Christy went to the Palomar Sky Atlas and checked but found that there weren't any sufficiently bright stars close enough to Pluto's position in April and May of 1978 to blend with the Pluto images and create the elongation. So it wasn't a background star unluckily nestled behind Pluto.

Something *was* up, and it was time to find his boss. Christy remembers that Harrington (see Figure 2.9) responded to the claim that the plates might be revealing a satellite of Pluto with one short sentence: "Jim, you're crazy."

The next morning, Friday, Christy still couldn't let go of the idea. So he went down to the Naval Observatory's collection of archival plates, which contained images of Pluto stretching back for decades. He found a set made in June of 1970, and damn it, the little elongation was not only there, it cycled around Pluto over a span of a week!

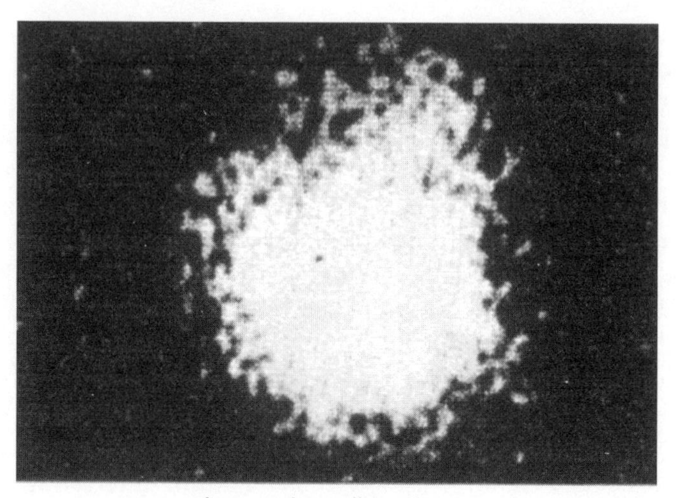

Pluto and Satellite Charon

FIGURE 2.8

The original 1978 photograph on which Charon was discovered by James Christy: It was taken with the U.S. Naval Observatory's 1.55-meter-diameter telescope in Flagstaff, Arizona, just 4 miles from where Pluto was discovered some 48 years earlier. On this image, Pluto is the large white blob, and Charon is the small extension the right of Pluto. Although the two are actually separated by more than 17,000 kilometers, blurring by the Earth's atmosphere made the two appear to be a single pear-shaped blob. Christy and Harrington's discovery of Charon was a breakthrough, as important as Cruikshank & Company's discovery of methane 2 years before. Charon's circular orbit over Pluto's equator has a period of 6.387 days, exactly equal to Pluto's periodic rotation period. This synchronization causes Charon to "hang" over one longitude on Pluto's surface. As a result, visitors to Charon would only "see" one-half of Pluto, and visitors on Pluto would not see Charon unless they were on the Charon-facing hemisphere. Although Charon would go through lunar-like lighting phases and move against the stars, from any given spot on Pluto, Charon would remain fixed in the sky, appearing to always hang in the same direction. (U.S. Naval Observatory)

FIGURE *2.9*

James Christy (seated), who discovered Charon, with his colleague Robert Harrington (ca. 1980): Harrington doubted the discovery at first but later became convinced Charon was real and used Charon's orbit to determine the mass of the Pluto–Charon system. (U.S. Naval Observatory)

This was so close to Pluto's rotation period that Christy figured it couldn't be a coincidence. So Christy assumed the satellite's orbit was precisely 6.387 days, Pluto's rotation period, and asked Harrington to calculate which side of Pluto the little elongation should lie on in each of the plates made in June of 1970, some 8 years and almost 500 rotation periods before. Nature would have to be perverse indeed to create a fluke of this magnitude. If this checked, the satellite probably was real.

According to MIT astronomer Rick Binzel, who was then working as a summer student at the Naval Observatory, Harrington and Christy worked independently to make this check, so that neither knew what answer the other found. When both had completed their work, they met in Harrington's office. "What did you get?" "Well, what did *you* get?"

As it turned out, Harrington's calculations of where the satellite should be agreed amazingly with Christy's measurements of the elongations seen on the 1970 plates. Less than 2 hours later, Christy found and checked two additional plates made back in 1965. Again the elongation appeared right where Harrington predicted it should be.

The satellite was real, and it was apparently in an orbit that was synchronized to Pluto's rotation period. Barely over a day had flashed by since Harrington had first sent Christy to routinely measure these possibly defective plates. But in that short span of time Christy had discovered that Pluto has a moon.

While other astronomers at the Naval Observatory in D.C. continued testing the satellite idea against various other alternatives, the USNO's Flagstaff station began a program to obtain more observations of the strongly suspected satellite. As they began, it still seemed somehow unreal: Pluto's moon had been found *entirely by accident*.

How could this be? Many times in the past, astronomers had searched for satellites of Pluto. Indeed, it was such an obvious thing to do that even Slipher, Lampland, and Tombaugh had searched for satellites back in March of 1930, the month they announced the discovery of Pluto. Later, in 1950, Humason (a solid observer of legendary repute, but the same man who had missed Pluto itself in 1919) conducted a dedicated search for satellites of Pluto. Others tried as well. Lowell Observatory's Otto Franz, who took the 1965 plates that Christy used to confirm the discovery, later said that he had noted the elongated image of Pluto in 1965 but dismissed it as a result of atmospheric turbulence. Franz hadn't noticed what Christy had—that only Pluto was elongated. As a result, Franz was still gently kicking himself over that one into the 1990s.

With hindsight, it's clear that Pluto's satellite (see Figures 2.10 and 2.11) wasn't such an easy thing to find. Why? For one thing, the satellite is 10 times fainter than Pluto, and it never strays farther than 1/4000th of a degree from Pluto as seen from Earth. Thus, the little moon is usually lost in the glare of Pluto. Further, as astronomer Bob Marcialis later described in a detailed analysis, several factors had conspired to keep the companion a secret for the 48 years since Pluto's discovery. Among these were the use of small telescopes in early satellite

FIGURE *2.10*

A modern ground-based image of Pluto and Charon: Though the two bodies are well resolved, no surface detail can be seen. The presence of Earth's atmosphere limits the capability of all telescopes on the surface of our planet. (Nordic Optical Telescope)

searches. In addition, past satellite searches took long exposures to detect faint objects relatively far from Pluto; such exposures had oversaturated the Pluto image and wiped out any chance of seeing a close-in satellite. Also, *any* putative moon was harder to see between the 1930s and the 1960s simply because of Pluto's greater distance from the Sun.

Once Harrington and Christy (and other astronomers at the Naval Observatory) were convinced that Pluto's satellite was real, it fell to Jim

FIGURE *2.11*

A Hubble Space Telescope image of Pluto and Charon taken in February 1994: This image clearly resolves the two objects as separate disks, but does not show surface detail. (R. Albrecht, ESA/ESTEC, and NASA)

Christy to name his find. Within a day of his discovery, he offered his wife, Charlene, the honor: "I could name it after you, Char-on." He was thinking it rhymed with prot-on and neutr-on. Another romantic astronomer.

To make a very long story short, however, Christy realized within a few days that although it was his privilege to propose a name, the rules of the International Astronomical Union's (IAU) nomenclature bylaws within which he operated stated that he would have to choose from Greek or Roman mythology. Someone suggested Persephone, the wife of Pluto. Christy liked that, so he went to a dictionary to get himself out of the trouble this could cause with his wife. But to his amazement, there, in black and white on the page, was the ancient name "Charon." In Greek mythology, Charon was the repugnant boatman who rowed dead souls over the river Styx into Hades, where the god of the underworld, Pluto, ruled. Christy realized this was the solution to his dilemma with both the IAU and his wife. Although correctly pronounced "Khar-on," Christy pronounced it "Shar-on," like Charlene—who promptly got her moon!

It was done. Pluto's satellite, which had for billions of years remained nameless in its anonymity, had acquired a name less than a week after its discovery by one very careful *Homo sapiens*.[8]

CHARON'S HARVEST

Charon's discovery was announced on July 7, 1978. If Dale Cruikshank fired the Pluto shot heard 'round the world with the discovery of methane on Pluto's surface, then Jim Christy's discovery of Charon was the second volley. Pluto was getting more interesting by leaps and bounds.

Within a few weeks, observations to refine Charon's orbit were pouring in from observatories around the world, and the basic orbit was determined to be a simple circle, without any noticeable ellipticity, at a distance somewhere between 17,000 and 20,000 kilometers (10,000 to 13,000 miles) from Pluto's surface. Because orbital mechanics dictates that satellites orbit over a planet's equator, Charon's north–south orbit on the sky meant that Pluto's equatorial plane also

[8]Actually, although the name Charon was widely used, it did not become officially recognized by the IAU until 1985 when the onset of mutual eclipses removed the last doubt that Charon was a satellite, rather than a mountain on Pluto.

pointed north–south, confirming Anderson and Fix's finding that Pluto's polar axis was tipped at least 50 degrees. In fact, using Charon's orbit as a plumb line, Pluto's obliquity was determined to be about 120 degrees, even more than Uranus's!

Using only the 6.387-day orbit period and the roughly 19,000-kilometer (about 12,000-mile) radius of Charon's orbit, Kepler's third law allowed astronomers to calculate the combined mass of Pluto and Charon. Skipping the arithmetic and cutting to the chase, Pluto's mass was immediately estimated, and it was even tinier than had been expected—a paltry 2 thousandths of the Earth's mass! That meant that Pluto's diameter was likely to be between 1,920 and 3,800 miles, much as Cruikshank, Pilcher, and Morrison had inferred on the basis of a bright, methane coating.

Christy's discovery also halted any chance that Pluto's mass would fall to zero, as Dessler and Russell had japed, but the catch was very close. Compared to both the predicted mass of Pluto and to *every other* planet, Pluto was, well, a miniature.

If a mass of 2/1,000 that of the Earth doesn't quite grab you, then consider this: Planet Pluto's mass is less than a tenth that of our own Moon. Beside its giant planet neighbors, Pluto isn't just a dwarf among giants: it's a flea.

Based on Charon's brightness, which was about one-sixth that of Pluto, it was immediately estimated that if the pair had similar reflectivity and internal density (a reasonable first assumption), then Charon's mass had to be in the neighborhood of one-tenth of Pluto's, and the satellite was likely to be between one-third and two-thirds the size of Pluto.

Nothing like this pair had ever been seen before in the solar system. The diameter of most satellites are far smaller than 1% of the diameters of their parent planets. Even Luna in the Earth–Moon system, the previous record holder for comparable size to its parent, has a diameter just a quarter the diameter of the Earth. Charon, however, was 50% the diameter of Pluto.

Put another way, whereas roughly 50 Lunas could fit within the volume of the Earth, only about eight Charons could fit within Pluto. This in turn meant that the balance point, or *barycenter,* of the planet and its moon lies between the two in open space—which is unique in the solar system. In every *other* case, the planet is so massive compared with its satellites that the system barycenter lies deep inside the planet itself. Thus, in comparison to all the primary-planet–puny-satellite

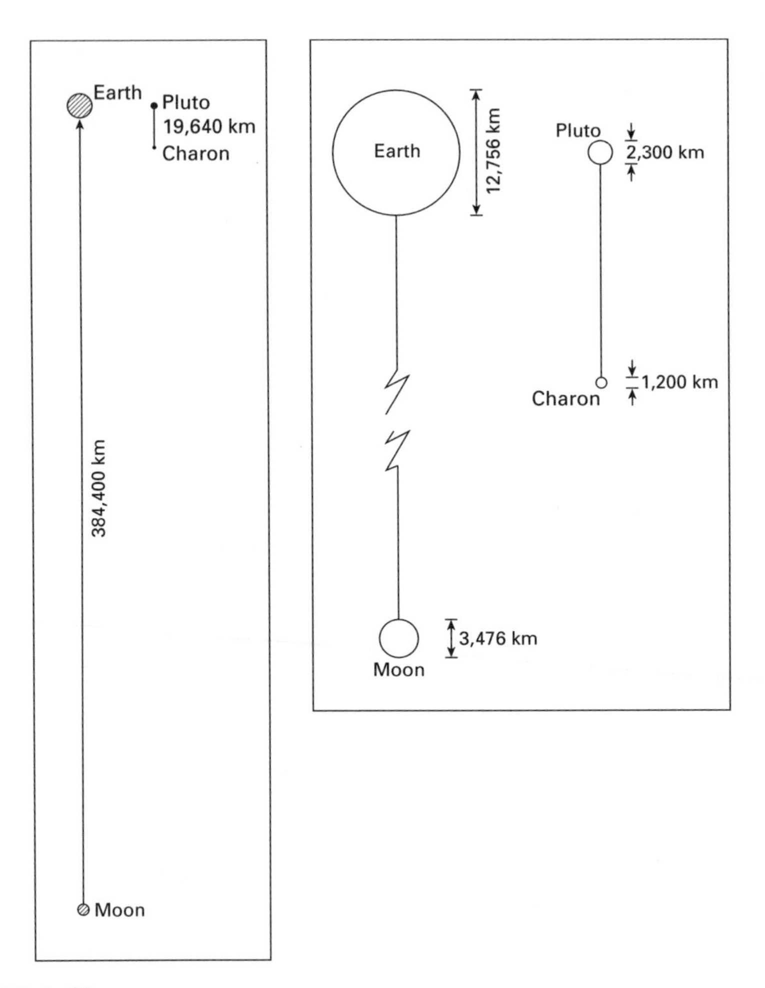

FIGURE *2.12*

Left: The Earth–Moon and Pluto–Charon systems to the same scale—Charon is a barely perceptible dot. Right: the same, with the planets and their moons scaled up by a factor of six—had we not broken the scale for the Earth–Moon system here, the Moon would be about a meter (slightly over 3 feet) off the bottom edge of the page.

pairs in the solar system, Pluto and Charon turn out to be nearly twinned in diameter, with the satellite half the size of the planet! When this was realized, Pluto–Charon became known as the first true binary planet system, as shown in Figure 2.12.

CHAPTER

3

A DISTANT DANCE

Then I felt like some
watcher of the skies
When a new planet swims
into his ken
KEATS

Christy and Harrington's discovery of Charon opened a new chapter in the study of Pluto. It wasn't just that the discovery was wholly unexpected, or that Charon attracted new attention, though it did. Nor was it the simple, almost fragrant freshness of the find, for satellites of planets had been discovered scores of times before. It was something different—something with more impact than the routine finding of a moon whizzing 'round a distant planet. It was that the long-abstract concept of Pluto—that pinpoint of light so far away—gained a new kind of reality. With Charon by its side, Pluto was no longer just the featureless dot it had been for so many decades. Those little, elongated images of the Pluto–Charon *system* made Pluto more real and tangible by giving it a dimension and, with that, a reality it had previously lacked. Thanks to Charon, Pluto left the long list of anonymous embers drifting far against the Deep, and joined the other planets as places the eye remembered as a destination, with extent, and shape, and just perhaps—a face.

THE GIFT OF A SWEDE

Everyone has a favorite planet, and to Leif Andersson, that was Pluto. Andersson was a Swede of little means, who so badly wanted to attend college that he financed it by entering and winning Swedish TV game shows. After undergraduate school in Sweden, Andersson went to Indiana University in the United States to earn a Ph.D. studying astronomy. There, he became fascinated with Pluto, though no one we spoke to quite remembers why. His Ph.D. advisor, Martin Burkhead, recalls Andersson back in the early 1970s as "an incredibly talented student, scientifically independent, and likely to become one of the leaders of his generation of planetary scientists."

Leif Andersson chose to base his doctoral dissertation on accurate photometry of the surface properties of the icy satellites of the outer planets. Despite the fact that Pluto was not an icy satellite, Andersson included it on his list of targets. How could he not? Somehow, in those early days before the depth of Pluto's mysteries were even clear, this Swede had taken little planet Pluto to his heart. Leif had become a Plutophile, long before being a Plutophile became cool.

Leif Andersson finished his doctoral research in 1975 and took a position as a postdoctoral researcher at the Lunar and Planetary Laboratory in Tucson, Arizona. There, just weeks after Charon was discovered in 1978, Leif was the first to realize that Charon was to become the key to unlock the door guarding many of Pluto's secrets.

We don't know exactly how or where Andersson got the idea, but sometime in the summer of 1978, a revelation washed over him: Pluto and Charon were going to undergo a set of *occultations* (i.e., events in which they would be moving in front of, or blocking, one another); this would give astronomers a fantastically sensitive probe of the Pluto–Charon system far beyond what they could otherwise have seen with the instruments of the day. If Andersson's calculations were right, these *mutual events* would yield precise information on the sizes, densities, compositions, and even the surface markings on Pluto and Charon. The information contained in the lightcurves of those mutual events would be manna from heaven for planetary astronomers.

The crux of Andersson's realization was elegant in its simplicity. As Pluto orbits the Sun, the direction of its polar axis remains almost precisely fixed in space (just as the Earth's polar axis is pointed in space toward the north star, Polaris). This *axial stability,* which is common to all of the planets, is generated by the great momentum inherent in any large object that spins. Although the pole of a planet can wander slowly over time, over short periods—hundreds of years—it hardly moves at all. Pluto's pole, for example, has been aimed toward a point between the bright stars Spica and Antares since long before recorded human history. This stability creates the interesting consequence: we here on Earth, deep down toward the Sun near one focus of Pluto's orbit, see Pluto's pose (which astronomers call its "aspect") from different angles as Pluto moves around the Sun (see Figure 3.1). If Pluto weren't tipped so severely on its side this change wouldn't be so dramatic. However, because Pluto *is* tipped over (something John Fix and his team had discovered early in the 1970s), it displays to us a dramatic range of aspects during each 248-year orbit.

What Leif Andersson realized before anyone else, is that as Pluto's aspect changes, Charon's orbit (which is locked into Pluto's equatorial plane and therefore changes aspect with Pluto) will be seen from various angles (see Figure 3.2). Sometimes, it will appear to circle Pluto like the ring around a bull's-eye, and at other times, when we see the orbit more edge on, the circle we see will collapse to an ellipse, and then, briefly, to a line when it lies completely edge-on as seen from Earth. Charon's orbit, mind you, isn't changing, but the angle from which we see it is. The key consequence of this change in perspective, Andersson recognized, is that twice each 248-year Pluto orbit, precisely 124 years apart (half the orbit period), Charon and Pluto undergo a few years when each repeatedly passes directly in front of and then behind the other, as seen from Earth, creating a series of scientifically valuable occultations.

Andersson estimated that each of these occultations would last for hours as each object would pass in front of the other during Charon's 6.387-day orbit around Pluto. At first, the events would put just a sliver of each object in darkness, as the shadow of one clipped the *limb* (outer edge) of the other. Then, over a period of 3 years, as the alignment perfected itself, the plane of Charon's orbit would cut longer and longer traverses across Pluto until, finally, Charon would transit the diameter of Pluto in what astronomers call "central events." In a *central* event, we on Earth would see the whole of Charon's disk (and its shadow) moving across Pluto's face for a few hours. Likewise, when Charon moved behind Pluto's disk, Charon would be completely engulfed and hidden for a few hours.

Based on the crude size estimates for Pluto and Charon that were then available, and the altitude of Charon's orbit, Andersson predicted that the prized central events would take place for about 2 years in the middle of the series of mutual events. After that, the changing relationship between Pluto's position and the Earth and Sun would begin to make Charon's orbit appear less well-aligned, and therefore make the

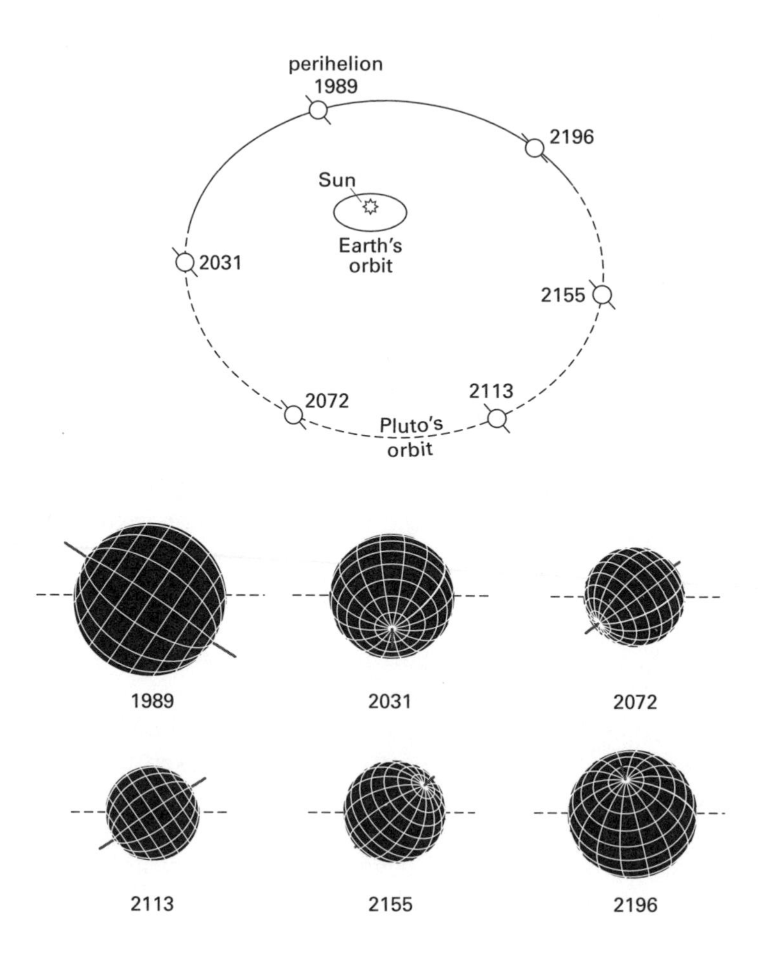

FIGURE 3.1

Pluto's rotation axis is tilted at an angle of 122 degrees from the plane of its orbit, making Pluto appear to lie on its side! As with any planet, Pluto's spin axis stays pointing to the same direction in space (top) as it orbits the sun. However, as Pluto progresses around its orbit, the view of Pluto from Earth—its aspect—gradually changes (bottom). These schematic diagrams (not to scale) show Pluto's location in its orbit at intervals through the 248 Earth years of a Plutonian year, along with the corresponding views of Pluto from Earth. Note also how the apparent size of Pluto varies with its changing distance from Earth during the orbit. The section of Pluto's orbit shown with dashes is the portion of the orbit that lies south of the plane of Earth's orbit.

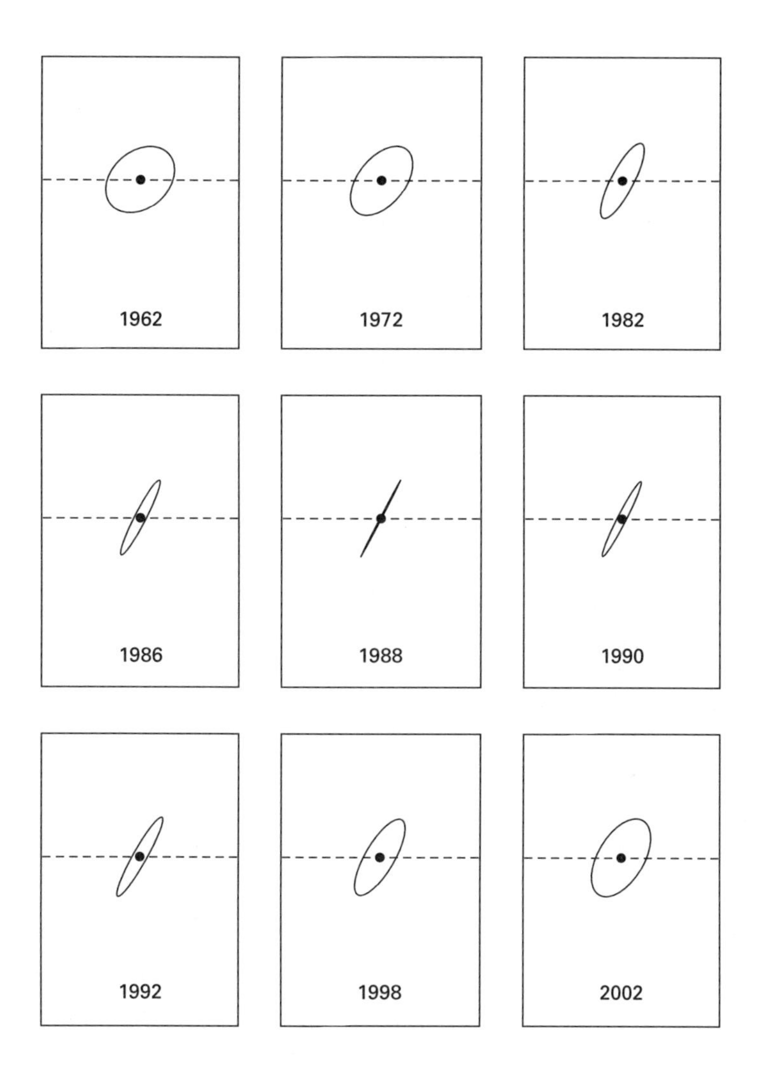

FIGURE *3.2*

The perspective from which we see Charon's orbit around Pluto gradually changes as Pluto travels around the Sun. This series of diagrams shows the orientation of Charon's orbit, as seen from Earth at various dates between 1962 and 2002. The tiny (<2 arcsec wide) ellipse that Charon's orbit sweeps out on our sky will diminish in size as Pluto's distance from the Sun increases during the course of its orbit.

occultations shorter and shorter in duration until about 3 years after the last central event, the last grazing crossing would occur, and the long waltz would end. All told, the Pluto–Charon occultation "season," as astronomers called it, would last between 5 and 6 years.

Based on the results Fix & Co. had published early in the 1970s, Andersson knew roughly where Pluto's pole was pointed and what its approximate orbit was. From that, he calculated something truly amazing: The events would occur in the 1970s or 1980s! Because no one was sure whether Pluto was rotating forward or backward about its axis, it wasn't mathematically certain whether the events were just finished or just about to begin. However, the fact that no observer had accidentally stumbled onto one in progress hinted it was more likely that the events were in the future, than in the recent past.

What luck! Each Pluto–Charon occultation season lasts only 5 years or so and occurs only twice each 248-year Pluto orbit. Andersson's calculations could have found that these rare events would next take place any time in the next 124 years (half of Pluto's orbital period). Andersson could well have discovered that the mutual occultations and eclipses between Pluto and Charon had last taken place virtually any time during the 6 long decades between Pluto's discovery and Charon's, and that we had missed them. He could have found that they occurred even farther back, while Lowell and Pickering were searching for Pluto, or before that, when the great men were but boys. But instead his numbers indicated that the events would begin *almost immediately.* Sometimes nature provides a gift, and this was one of those times. By sheer luck, Charon had been discovered about 120 years after its last shadow dance with Pluto and, therefore, on the eve of a new set of such events, pregnant with scientific promise.

Like many pregnancies, those around the expectant couple didn't know exactly when the due date was. Why? Because both the direction of Pluto's polar axis and the inclination of Charon's orbit were each uncertain to some degree. Neither Andersson, nor anyone else, could predict exactly which year the events would begin. Andersson's first calculations indicated a start in 1979 or 1980, but new orbit-inclination and pole-direction refinements, motivated by the promise of the impending events, soon revealed that they probably wouldn't begin until at least 1982, and perhaps not until 1985. Still better predictions weren't in the cards, owing to the uncertainties inherent in the data fed into the calculations, and so it fell to the observers to laboriously hunt down the events, searching year by year until the first one was bagged. Once any one event was found, it would be possible to predict future events fairly accurately, since they would occur at 3.194-day intervals—exactly half a Charon orbit apart.

THE DANCE OF THE DUET, DISSECTED

What would it be like to be there, on Pluto during one of the mutual events? First, Charon's 4-degree-wide disk would approach and then move over the bright but tiny, distant Sun. As it did, Charon's shadow would race across Pluto's frozen terrain at roughly Charon's orbital speed, some 220 meters per second (about 730 feet per second). During such an event, Pluto's snowy surface would be lit only by starlight reflected off the reddish annulus of Pluto that remained out of Charon's shadow. After a few minutes or hours of darkness (depending on the location on Pluto's surface and how deep into the 6-year event season it was), the Sun would then reemerge to reassert its diluted light and feeble warmth.

As seen from the warm confines here on terra firma, the same event would slowly unfold through four phases. It would begin with *first contact,* when either the occulter or its shadow (depending on the angle at which the event is seen from Earth) would first impinge on the occultee. *Second contact* would occur when the occulter and its shadow lie entirely on the occultee. *Third contact* would then occur when the occulter or its shadow began exiting the occultee. Finally, the event would end with *fourth contact,* when the last slice of the occulter or its shadow would leave the occultee. The period of *centrality* would be the interval between second and third contact, when Charon and its shadow lay entirely on Pluto (or, in a Charon occultation, when Charon was entirely hidden by Pluto). Of course, during the first and last years of the mutual-event season, when the events were grazing rather than central, only the first and fourth contacts occurred.

If the Hubble Space Telescope (HST) had been flying back in the early 1980s, it would have been possible to actually watch the progress of each mutual event, as the two bodies seemed to merge, to see the shadow of the leader fall on its rearward partner, scoot across the eclipsee's surface, and then make its exit of the eclipsee's limb. But HST remained an unfinished work, and no Earth-bound telescope could then overcome the effects of our blurring atmosphere to resolve Pluto (or, for that matter, even reliably split Pluto's image from Charon).

As a result, astronomers were reduced to their tried and true technique of making lightcurves, graphs of the brightness of the system, in order to detect and study the mutual events, as shown in Figure 3.3,

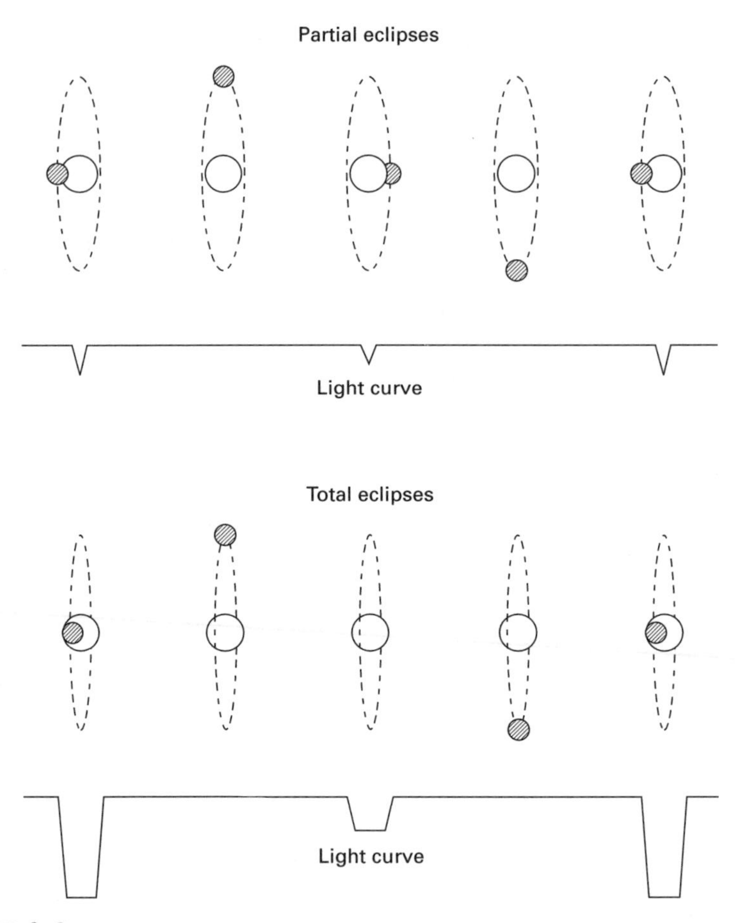

FIGURE *3.3*

A schematic representation of the shape of Pluto and Charon's lightcurve for mutual events involving partial eclipse (top) and total eclipse (bottom): The sketches of the mutual event phenomena are not to scale.

simple lightcurves using either PMTs, or their CCD cameras, were the tools used to probe the mutual events.

CCD cameras are the miniaturized, modern-day digital equivalent of photographic cameras. They came into wide astronomical use in the 1980s because they combine the best attributes of the PMT (i.e., digital nature and strict linearity) with the best attribute of photographic plates: A CCD produces an image. CCDs create their digital images from millions of tiny picture elements, *pixels,* each far smaller than a grain of sand. In effect, a CCD produces the same result as would millions of close-packed PMTs, but at a far lower cost, and in a space only an inch or so on a side.

And what would the PMTs or CCDs see when trained on one of the events? The schematic event lightcurves shown in Figure 3.3 illustrate how each event causes a decline in the total brightness of the pair (because part of one body is blocked from view). The decline begins at first contact, progressively increases until the event reaches its deepest stage, and then reverses itself as the event wanes. For the first events, it was predicted that the eclipsed sliver would be so small that the total decline in brightness would be hard to detect—perhaps only 1% or 2% and that would have to be detected against the lightcurve due to Pluto's rotation.

Within months of Leif Andersson's prediction, the power and potential of the onrushing event season became clear, and dozens of astronomers began to organize observing plans. Tragically, however, Leif Andersson was denied the opportunity to take part in the very events he himself had forecast. Leif contracted and quickly succumbed to lymphoma in 1979, at the age of 35. Had he lived, he would probably have been one of the central characters of the pages that follow. Instead, he is remembered only for launching the greatest single Pluto research bonanza of all time.

INVITATION TO A DANCE

By the time the calendar rolled into 1982, groups of astronomers in Hawaii, Arizona, California, Texas, Germany, the Soviet Union, and South Africa had planned extensive observational campaigns to detect the onset of the Pluto–Charon mutual events. These campaigns weren't easy undertakings. Not only were predictions of the month and year of the first events uncertain, but uncertainties in Charon's precise orbit meant that each possible event had its own inherent uncertainty of up to a few hours (equivalent to roughly 2% of a Charon rotation period). As a result, the observers not only had to face the fact that they would probably have to observe dozens of candidate events over several years before they might snag one, but also the fact that each event could occur within a wide window of time on a given night. Patience became a virtue newly appreciated.

Despite the hard work of several groups, no events were detected in 1982, or 1983, or early 1984. Each year, there had been more than 100 potential events, about a third of which would occur during darkness at one of the observatories where patient people waited. Long nights passed by the bucketload. Still, no one observed even the slightest hint or the subtlest signature of the first stages of the oncoming ballet between Pluto and Charon.

When 1984 ended without reward, some astronomers were beginning to wonder if the events would really take place at all. Prediction calculations were rechecked: indeed—the events should now begin very soon, but when? When? When?

TALLY HO SHADOW!

Bonnie Buratti, a young astronomer with a new job at the Jet Propulsion Laboratory (JPL), was the first to spy something. Early on the morning of January 16th, 1985, Buratti was observing Pluto with Ed Tedesco at Mount Palomar, high above the LA basin. Tedesco had long been involved in the search for the first Pluto–Charon events and had asked Buratti to help with some of his observing. As they reported it, they seemed to have stumbled onto Charon crossing over Pluto in the early, predawn sky fully three hours before the nominal time of the predicted event for that day.

What Tedesco and Buratti's data contained was a subtle signature. But this was almost too subtle: just three lightcurve data points, each depressed only a few percent from what they normally would have been.

Was this the real thing, or was it something else? Tedesco and Buratti had checked the brightness of a pair of comparison stars in between each Pluto measurement—and the standard stars hadn't changed in intensity the way Pluto had. This argued that the signature they saw in their data could be real. However, the fact that the event occurred 3 hours earlier than expected was worrisome. It was widely known that the exact timing of predicted events could be off by an hour or more, but the probability that an event would occur this far off the mark was less than 1 in 1,000. Equally if not more troubling, was the concern that Buratti and Tedesco's CCD had been running low on coolant just as the shallow, three-point signature occurred, suggesting the possibility that the whole thing was simply a slight instrumental change, rather than a change in Pluto. All told, the result was a murky, frustrating kind of thing that begged for corroboration. However, as luck would have it, no one else had planned and reserved telescope time to observe the January 16th event, so there wasn't a second data set to check against. Without more data, no one could be sure whether the event had been real.

Buratti and Tedesco passed the news of their "event" to other mutual-event watchers, using the primitive but burgeoning scientific e-mail networks of the time. Buratti and Tedesco noted their concern about the warming of their CCD and the event's early occurrence, but all the same, they pointed toward their data as tentative evidence that a mutual event might finally have been detected.

Murphy's laws, of course, always apply in situations such as this. No one was scheduled to observe another potential event for about a month. Why? Many of the succeeding events would occur in daylight, and some would occur with Pluto too close to the Moon to be observable.

So salvation had to wait until February 17th, when a young and exceptionally talented, Ph.D. student named Richard "Rick" Binzel, who was working at the University of Texas, was scheduled to observe another predicted event. Binzel mounted a PMT on the 0.9-meter (36-inch) diameter telescope located at McDonald Observatory in the Davis Mountains of West Texas, and caught an entire, 2-hour mutual event through crisp, clear west Texas skies.

Binzel had started observing Pluto as early as possible that night and captured the entire event from start to finish. He analyzed his results as they came in, on a programmable hand calculator.[9] The data showed a smooth decline and then a symmetric rise in Pluto's brightness that looked right. But Binzel wanted to be sure, so he rechecked his calculations twice. By 9 A.M., he was convinced he had something real and started making phone calls.

The person Binzel informed first was his Ph.D. advisor, the director of McDonald Observatory, Harlan Smith. From his home in Austin that Sunday morning, Smith asked Binzel, "Are you *sure*?" Binzel said yes, and Smith didn't challenge him. Smith trusted Binzel's talent and his word. Realizing the import of the news that the mutual-event season was finally on, Smith replied, "OK, we'll do a press release." Binzel then called Charon's discoverer Jim Christy to tell him the exciting news personally. That done, he began sending e-mail to some of the other observers involved in the search for the first mutual events.

[9]Binzel's event had also occurred quite early, apparently in agreement with Tedesco and Buratti's surprisingly early event. Years later, however, when many events had been observed, calculations showed that the Tedesco and Buratti's "event" very probably occurred outside the time that an actual event was observable. Although most mutual-event experts now believe the January 16 report was a false alarm, Tedesco and Buratti's contribution to the first detection of mutual events remains an important one.

Within a day, the news was public, wafting through newspapers across the world.

Then exactly—half a Charon-orbit later—3.194 days after Binzel's first event, on February 20th, another young astronomer, the University of Hawaii's David Tholen, caught Charon falling in Pluto's shadow, using a PMT on the 2.24-meter (88-inch) University of Hawaii telescope on Mauna Kea in Hawaii. That clinched it. The dance had begun! See Figure 3.4.

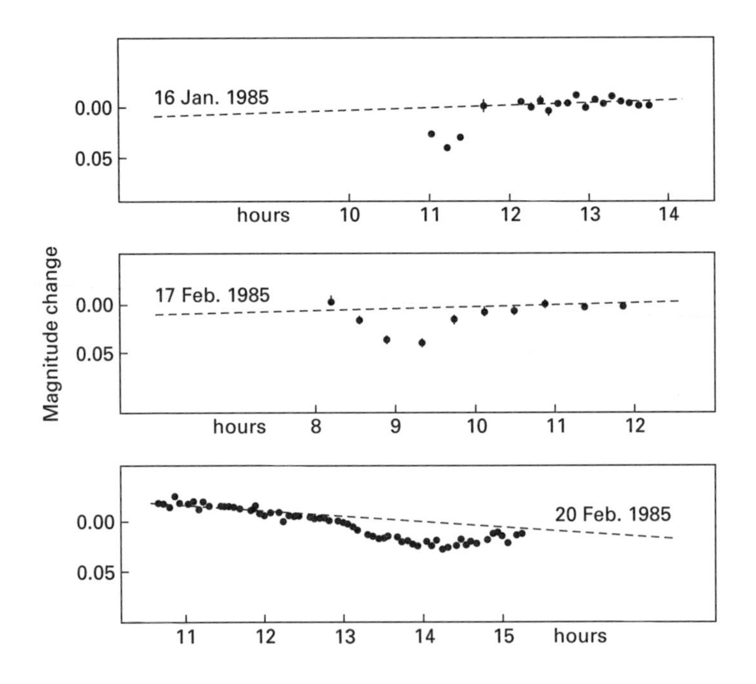

FIGURE *3.4*

Observations of the first three mutual-event lightcurves observed, back in 1985: The black dots are the measurements of the change in the combined brightness of Pluto and Charon together, plotted as time sequences during the events. The dashed line shows what was expected had there been no events; the string of dots dips below the dashed line during each event. The time is given in Universal Time (UT) for all three dates. These three lightcurves have been aligned one above another, in relation to Charon's position in its orbit. This shows that the dips are linked to the relative position of Pluto and Charon and are therefore unlikely to be caused by anything other than the anticipated mutual events. The January 16, 1985, observation was made by Ed Tedesco and Bonnie Buratti at the 1.5-m reflector at Palomar Observatory. The February 17 event was recorded by Richard Binzel at the McDonald Observatory; the February 20 observation was made by David Tholen on Mauna Kea, Hawaii. (Adapted from Binzel et al., *Science,* June 1985)

GLIMPSING THE TREASURE TROVE

Within weeks, Binzel, Tholen, Tedesco, Buratti, and a JPL colleague, Bob Nelson, had teamed together to write up their findings and submit them to the prestigious scientific journal *Science*. In this report, Binzel and company reached two important conclusions from the data sets they had gathered from the first mutual events.

The first result followed from the finding that the depths of the January 16 and February 17 event lightcurves, in which Charon partially obscured Pluto, were significantly greater than the depth of Tholen's February 20 lightcurve, in which Pluto partially obscured Charon. From that observation, the five scientists concluded that the terrain on Pluto that had been eclipsed by Charon's shadow must be more reflective than the terrain eclipsed on Charon by Pluto's shadow. This was the first piece of evidence foreshadowing just how strikingly different Pluto and Charon actually are. There would be much more evidence to this effect from later events.

The second result the five researchers obtained came from the fact that the difference in time between Pluto's eclipse of Charon and Charon's eclipse of Pluto was within minutes of the 3.194 days, a half-orbit of Charon. The fact that the timing was so exact meant that Charon's orbit wasn't just roughly circular, but was damn near exactly circular, perhaps not deviating by more than a few kilometers over its nearly 40,000-kilometer span!

The starting gun had been fired, and the Pluto–Charon mutual events were yielding their first quantitative results. Over the next 6 years, as the events progressed from the shallow graze first detected in 1985 to the central events of 1987 and 1988, and then back, toward the final grazes that occurred in 1991, observers around the world became involved in the careful prediction of new events, and their observation. Others also became involved in the synthesis of data embedded in the lightcurves, racing across space from the frozen Plutonian wilderness 5 billion kilometers (3 billion miles) away. In all, more than 20 telescopes and 40 scientists ultimately became involved in this work, which yielded a bonanza of new insights into the solar system's distant, dynamic duo. Let's see what they learned.

WHEN SECONDS EQUAL CENTIMETERS: MEASURING PLUTO'S TRUE SIZE

Astronomers who point their telescopes skyward to study stars and planets have some important things in common with the high-tech spies who point their telescopes downward from Earth orbit. In both cases, the observer can look but not touch. So, too, both kinds of observers must often resort to wit and clever device to substitute for the limitations of their instrumentation and their perch. For exactly this reason, astronomers have long valued the power of mutual events between binary stars as a tool for leveraging limited telescopic resolution. So when planetary astronomers first learned of Andersson's prediction that Pluto and Charon would eclipse one another in the 1980s, the salivating was reminiscent of Pavlov's dogs.

As we described in Chapter 2, Pluto's size and great distance conspire to make its apparent size on the sky as seen from Earth truly minuscule—just 1/10 of an arcsecond (1/36,000 of a degree). This is something like the size of a grain of sand viewed from a distance of about a kilometer (1,100 yards)—far too small for even the best ground-based telescopes of the 1980s to resolve through the turbulence of the Earth's atmosphere.[10]

What a frustration! Fifty years after Pluto's discovery, its diameter had still not been directly measured, so its size still wasn't well determined. In fact, Pluto stood alone among the planets in this regard. Every planet was either close enough, or large enough, or both, to be easy to resolve in a telescope and therefore had a well-known diameter, except Pluto. How big was Pluto?

The best available information on Pluto's size before the mutual event season came from a calculation. Using the mass determination that resulted from Charon's discovery, astronomers could estimate Pluto's size, once they knew Pluto's *density*—that is, how much mass was packed into each cubic centimeter (or inch). Although the density of the ices and rocks that make up solid planets only varies over a limited range, the fact that Pluto's density was completely unknown meant that astronomers couldn't constrain its diameter very well. Pluto might be as small as 1,900 kilometers (1,100 miles) or as large as 4,300 kilometers (2,700 miles), but that was about it.

Amazingly, back in the early 1980s, Charon's diameter was better known than Pluto's. Why? For years, astronomers had tried to catch Pluto in the act of occulting (i.e., blocking) the light of a star, so that its size could be determined from the length of time the star was blocked and its rate of travel across the star—both of which are easily

[10]Recall that except in the best observing conditions, the best telescopes couldn't even cleanly split Pluto from Charon, a full arcsecond apart in the 1980s.

measured. Several such events had been predicted to occur, but Pluto never actually cut across the various stars astronomers hoped it might. The damn thing kept missing by a maddening few 10,000ths of a degree or so!

However, during one such widely anticipated event, back in April of 1980, South African astronomer Alistair Walker caught Charon occulting the star Pluto should have snagged. Once Walker's lightcurve of the occultation was analyzed, it became clear that he had, by good fortune, snagged a nearly central occultation by Charon. This allowed him to estimate Charon's diameter very accurately. Knowing that his photometer showed that Charon had blocked the star for 50 seconds, Walker was able to determine that Charon's diameter was almost exactly 1,200 kilometers (750 miles).

The mutual events would allow Pluto's size to be determined by turning the difficult (indeed, then-insurmountable) problem of resolving Pluto's size in a telescope, into a much easier problem—one of timing an eclipse between Pluto and Charon. Then, by measuring the interval between first and third contact as Charon glided over the surface of Pluto, and knowing Charon's orbital speed (about 100 meters per second), astronomers could quickly calculate the distance Charon had traveled across Pluto. For example, if the time from first contact to fourth was precisely 3 hours, then Charon's well-established orbital speed implies that the distance traveled was 2,400 kilometers (1,500 miles). Voilá! No wonder so many astronomers competed to observe the events. It's almost simple enough to be a scout's merit-badge project.

Well, maybe. It would take a goodly-sized telescope and a thorough knowledge of photomultiplier tubes and data reduction for the little scout to do the experiment. And he'd have to account for some pesky, real-life complications. For example, Charon's early shadow events were just grazing events across a short segment of Pluto's disk, rather than its full diameter. Therefore, a little scout would have to know Charon's orbit accurately enough to know which part of Pluto was being eclipsed or occulted (pole, equator, or precisely which latitude in between)—without that, you couldn't determine Pluto's diameter until all of the events were seen and the longest (across the diameter) was picked out. Maddeningly, of course, knowing which part of Pluto was being occulted by Charon, in turn depends in part on knowing Pluto's size, which is what one is trying to determine. There are other complications as well, such as shadows that cause the events to appear longer than they really are. That is, because the Sun and the Earth, as seen from Pluto are offset by up to two degrees, the line of sight from Earth can see over Charon's shoulder to the place where Charon's shadow falls on Pluto. As a result, the event lightcurve can start earlier or end later than the actual time of first and fourth contact because of shadowing. Better remember to put that in your calculation, and there's more, little scout. What if Pluto (or Charon for that matter) isn't perfectly round? Also, doesn't the angle from which we view the events change slightly during each event, as the Earth moves (at the modest clip of 100,000 kilometers [63,000 miles] per hour)? And what if Pluto has an atmosphere with hazes that might affect the lightcurves? . . . And so on. OK, a scout could take the data, if he had one heck of a telescope at his disposal, but his parents would probably have to help with data analysis.

We haven't run across any reports of scouts trying to measure Pluto's diameter for a merit badge, but we do know that several astronomers did. As it turned out, the single biggest headache (or challenge, depending on your perspective) in determining the diameters of Pluto and Charon from event timing was that the accuracy of the derived diameter depends directly on how well the size (semi-major axis) of Charon's orbit is known. By the mid-1980s, Charon's semi-major axis (which is the distance it orbits above Pluto's center) was known to be between 19,000 and 20,000 kilometers (roughly 11,800 and 12,500 miles). Still, that uncertainty meant that even if everything *else* that went into the calculation of Pluto's size were known perfectly, Pluto's radius would be uncertain at the five percent level.

In the 1990s, better data, much of it based on Hubble Space Telescope results, have allowed Charon's semi-major axis to be determined to within about 1.5% (results range from 19,400 kilometers [12,125 miles] to 19,700 kilometers [12,313 miles]). Combining these numbers with the timing data from mutual events gives a diameter for Pluto somewhere near 2,360 kilometers (about 1,475 miles), with uncertainties of just a few dozen kilometers (perhaps 40 miles). The cognoscenti are still arguing about the details. . . .

Still, the mutual events provided something that was both puzzling and laden with possible promise as a clue to Pluto's origin: Its diameter is far smaller than any other planet's. By comparison with Earth, for example, as shown in Figure 3.5, Pluto would not even stretch across the central United States! In fact, Pluto is so small that it's even smaller than Earth's Moon and six other satellites in the solar system: Ganymede, Callisto, Europa, Io, Titan, and Triton. Miniature indeed.

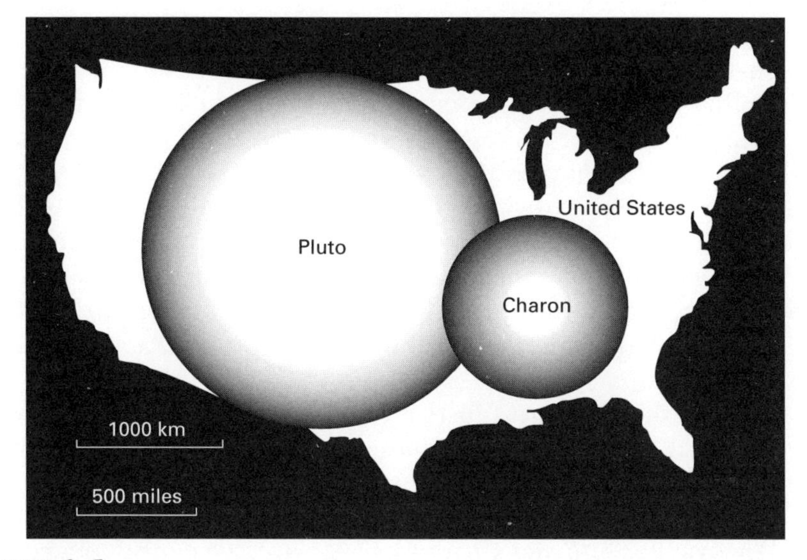

FIGURE *3.5*

The sizes of Pluto and Charon in relation to the United States: Recent estimates of Pluto's apparent diameter are near 2,360 km (1,475 miles). The best figure for Charon is near 1,200 kilometers (750 miles). Both estimates are thought to be accurate to within about 40 kilometers (25 miles).

BOUNTY

Knowing the sizes of Pluto and Charon opened up the door to other insights about the distant duo. One of the first things that the sizes of Pluto and Charon provided was the ability to translate the brightnesses of Pluto and Charon into estimates of their surface reflectivity, or *albedo*. This couldn't be done before their sizes were known because the albedo depends on how much reflecting area each world presents.

Using the sizes they determined from mutual-event data, Dave Tholen and Marc Buie found Pluto's average albedo is about 55% at visible wavelengths. Charon, by comparison, is far less reflective, with a visible albedo near 35%.

This reflectivity is fairly high compared with most large objects in the solar system. For comparison, the Moon's average reflectivity is about 12%, and the reflectivity of most of the icy satellites hovers around 35%, like Charon's. In fact, what this translates to is the implication that Pluto's surface is about as reflective as freshly fallen snow. And compared to the freshly driven Colorado ski-powder appearance of Pluto, Charon's surface might be thought of as week-old Boston snow.

Another finding derived directly from the mutual-event diameters was a measure of the *bulk density* (i.e., the average amount of stuff packed into each cubic centimeter) of Pluto and Charon. This quantity, which is computed by dividing an object's mass by its volume, is fundamental to making estimates of what a body is likely to be composed of. If, for example, an object's density is high, then it is likely to be composed of a good deal of rock; conversely, if an object's density is low, it is likely to contain a large fraction of lighter materials, such as ices, in its interior.

Knowing the total mass of the Pluto–Charon system by applying Kepler's third law to Charon's orbit (see Chapter 2) and the diameter sizes of Pluto and Charon from mutual event lightcurve timing, it was possible to calculate the average density of Pluto and Charon together.[11] The result: Pluto and Charon have an average density of about 2 grams per cubic centimeter—about twice that of water ice (at 1 gram per cubic centimeter), but significantly less than typical garden-variety rocks (at 3 grams per cubic centimeter).

The fact that Pluto and Charon have an average density near 2 grams per cubic centimeter means that the pair probably contain a mixture of ice and rock. Figure 3.6 compares the average density of Pluto–Charon to other worlds.

To constrain the interior composition more quantitatively, two teams—one led by NASA's Damon Simonelli and the other led by Washington University's Bill McKinnon—fed the average density into computer models describing the basic structural equations of planetary interiors. From these models, the two teams determined that the Pluto–Charon pair is probably composed of a mixture consisting of about 65% light, rocky minerals, and about 30% water ice (the most common ice in the solar system), with the remaining 5% or so comprising traces of heavier stuff (such as iron) and lighter stuff (for example, methane ice, which has a density about half that of water ice).

McKinnon took this result a little further in a later study with his collaborator Steve Mueller of Southern Methodist University in Dallas. What they wanted to know was whether Pluto's interior might have dif-

[11]Had astronomers known the separate masses of Pluto and Charon, then the individual densities of Pluto and Charon could have been calculated, but determining the individual masses of the pair was beyond 1980s capabilities. In fact, as of 1997, the individual masses still aren't well enough known to tell whether Pluto and Charon have the same or different densities.

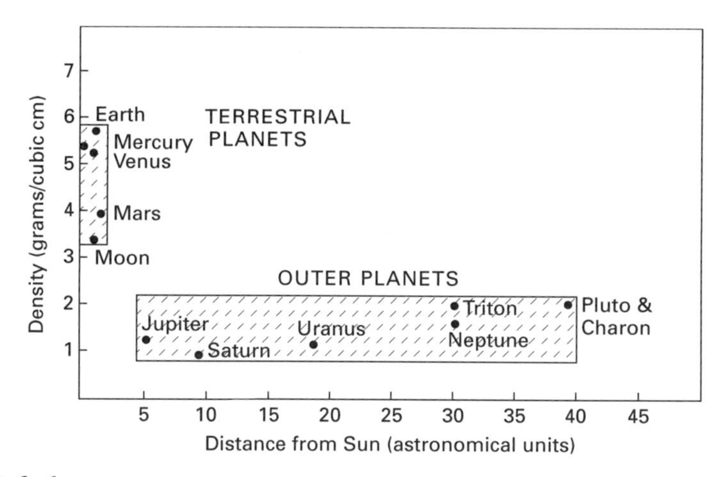

FIGURE *3.6*

The densities of the nine planets and a few other major objects are shown here in relation to their distances from the Sun. Note that there are two distinct groups: (1) the inner, dense, rocky planets, and (2) the outer, less dense giant planets, along with Pluto–Charon and Triton. (Courtesy, S. A. Stern)

ferentiated. *Differentiation* is the term used when a planet's interior is arranged into a set of layers, with the densest material at the center and the least dense material on its surface, rather than having a more or less uniform mixture of ice mixed with dirt and rock. Diagrams of such objects tend to resemble a kind of onion-skin arrangement (see Figure 3.7). Many of the inner planets and some of the large satellites of the outer planets have differentiated into a layered core–mantle–crust structure like that of the Earth. So, is it likely that Pluto's interior has distinct layers? The best test would be to fly a spacecraft by Pluto to measure the subtle difference in its trajectory that a layered versus a homogeneous (i.e., undifferentiated) Pluto would cause. Lacking that, McKinnon and Mueller ran computer models. What they found was . . . was . . . was . . . maybe. According to their calculations, Pluto's size and mass put it close to the borderline where planets differentiate under their own weight. To know for sure, according to McKinnon, we will likely just have to go to Pluto and find out.

GRAND CENTRAL

So, it wasn't a bad start. The little graphs of how the light level of Pluto plus Charon changed as one or the other passed in front of its partner had revealed the sizes and reflectivities of Pluto and Charon, and the

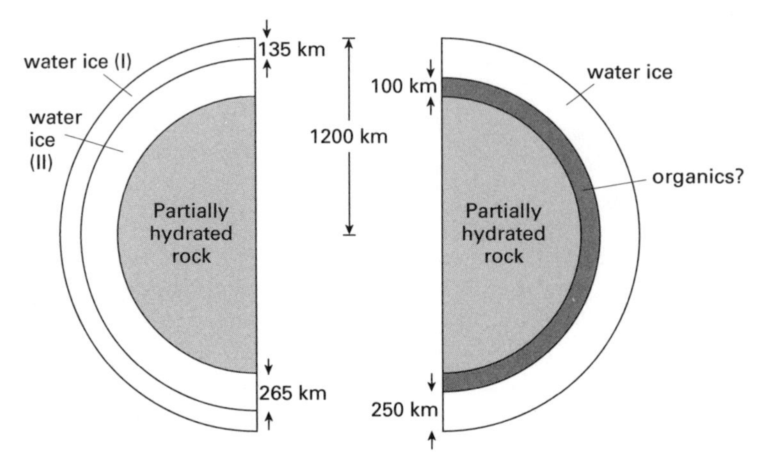

FIGURE *3.7*

Two possible models for the interior structure of Pluto: In both cases, the surface is covered by a thin layer of frosts, a few kilometers or tens of kilometers thick, which is not shown here. The model on the left is based on an average density on the low side of current-day estimates (1.735 grams/cubic centimeter [cm]), with 54% of the volume being rock and the rest ice. The lower part of the mantle in this case is a form of water ice (ice II) created under high pressure. The model on the right assumes a slightly higher density (1.85 grams/cubic cm) and a layer of organic material under the water-ice mantle. Owing to the higher density, the proportion of rock in this model is higher, at 70%. (Adapted from W. McKinnon et al., in *Pluto and Charon,* University of Arizona Press, 1997)

average density of the two bodies. For their next trick, the eclipse observers would use the very best events to reveal the separate surface compositions of Pluto and Charon.

They say that beggars can't be choosers—and indeed, the shimmering of the Earth's turbulent troposphere had, for centuries, made astronomers beggars when it came to achieving high resolution. If the atmosphere above us blends together the light from Pluto and Charon into some convoluted average of the two, then how could a ground-based observer cleanly split the pair to reveal the light from each individually? By the 1990s, there were new technologies to finally beat the atmosphere at its own game, but in the 1980s, this wasn't possible. For the beggars to become choosers, a little ingenuity would be required.

It shouldn't have taken a rocket scientist to realize this, but in fact it did. All anyone had to do was wait for a central event in which Charon was completely hidden behind Pluto. During those rare and precious periods, any telescope big enough to gather a good measurement would see Pluto alone, without any luminous contamination from Charon.

Bravo! Better yet, by taking measurements of the combined signal from the pair before or after the event, and then subtracting away the pure Pluto signal obtained when Charon is hidden behind Pluto and therefore out of the picture, astronomers could obtain the difference between "Pluto+Charon" and "Pluto only," thus revealing "Charon alone." Using exactly this trick, the beggars would become opportunists in the cause of exploration!

The race to unravel the separate compositions of Pluto and Charon thus came down to obtaining high-quality visible and infrared spectra between the second and third contacts of a central event (with Pluto eclipsing Charon). The main limitation in accomplishing this was that the Pluto-only spectrum had to be taken during the short, 2-hour interval when Charon had disappeared.

If the mutual events between Pluto and Charon had occurred in the 1970s, there wouldn't have been much hope for this kind of experiment because the available detectors just weren't sensitive enough to obtain a good spectrum of Pluto in the time available. However, by 1987, when the first central events completely hiding Charon were about to take place, the state-of-the-art capabilities for instrument sensitivity in the compositionally diagnostic infrared region had improved by a factor of 40 to 100 times over what Cruikshank and his colleagues had had at their disposal on Mauna Kea in 1976. With this kind of sensitivity and the onrushing opportunity to separate Pluto's composition from Charon's for the first time, planning for these observations began more than a year in advance. By the time of the first total disappearance of Charon visible at night, in March of 1987, the Plutophiles were out in force.

One observing group was led by graduate student Scott Sawyer at McDonald Observatory, in the Davis mountains almost 7,000 feet above west Texas. The same night, Marc Buie, by then a postdoc at the Institute for Astronomy in Honolulu, was ready on Mauna Kea, 14,000 feet above the shores of the big island of Hawaii. Meanwhile, Ed Tedesco and his colleague John Africano perched themselves atop Kitt Peak in Arizona to study the impending event.

But these three groups were not the only ones that set up shop. A careful, two-pronged attack was also planned by University of Arizona graduate student Bob Marcialis, who sat ready with professors George and Marcia Reike in the control room of the Multiple Mirror Telescope on Mount Hopkins, high above the southern Arizona desert. In addition, Marcialis was collaborating with University of Arizona professor Uwe Fink and his then-student, Mike DiSanti, who manned a telescope atop Mount Bigelow in central Arizona. Altogether, five groups had set up to capture the fleeting 4-hour-old photons showering Earth from the first observable central event. It was almost as if gold was falling from the sky.

Both Scott Sawyer's group and the Fink/DiSanti team were operating visible-wavelength spectrometers at the focus of their telescopes. These spectrometers would be used to determine how much of Cruikshank's methane signature was due to Charon. They would accomplish this by monitoring the change in one of methane's strongest spectral-absorption bands—near 0.89 microns—throughout the event. Buie, on Mauna Kea, and Marcialis on Mount Hopkins had each set up infrared PMTs designed to plumb the rich infrared region for what it might have to tell. The event was to begin in the early evening of March 14th, and would last several hours.

Twilight fell first on Texas that Wednesday evening, but clouds covered the sky, and a rain briefly fell. Downhearted, Sawyer and his coworkers at McDonald thought they would miss the disappearance of Charon behind Pluto altogether. But amazingly, the skies cleared shortly after first contact. All three Arizona teams (Tedesco/Africano, Marcialis et al., and Fink/DiSanti) enjoyed nearly perfect weather that evening, and all three obtained good data. On Mauna Kea, however, the winds were so high (120 mph!) that Buie had to close the telescope dome and give up without any data being taken. Although the skies were clear, Buie feared that staying open in the wind could damage the telescope.

It didn't take long for the various teams to reduce the highly anticipated data. In fact, the first results were officially announced just one week later at a scientific conference in Tucson, when Bob Marcialis took the podium on behalf of both his team and the Fink/DiSanti team. Marcialis (see Figure 3.8) is a short, stocky man with deep passions for both softball and science. That day, he swung his research bat three times, making solid contact at each swing.

When he showed the data contained in Figure 3.9, Marcialis told the hushed listeners that he and his coworkers had found that the methane absorptions first detected by Cruikshank and friends originated entirely on Pluto's surface. This was clear because the strength of

FIGURE *3.8*

Bob Marcialis: With coworkers, he discovered that the signature of methane frost from the Pluto–Charon system is due only to Pluto, while Charon shows evidence for the presence of water ice. Marcialis was also the first to work out a pattern of surface markings on Pluto that would be consistent with its rotational lightcurve. (Photo: Maria Schuchardt)

the methane absorptions had not weakened when Charon disappeared. The data also revealed that the visible-wavelength spectrum of Charon obtained by Fink and DiSanti was both featureless and without any perceptible color of its own. Unlike the reddish reflectance of Pluto, the color of the terrain observed on Charon was a pallid gray. Sawyer and his team would later report the same.

Finally, Marcialis let go his home-run hit: By subtracting the Pluto-only spectrum from the Pluto+Charon spectrum, the infrared data that he and the Riekes had obtained on Mount Hopkins revealed clear evidence for absorption bands characteristic of something never before seen in a spectrum of the Pluto system: water ice, and it was on Charon.

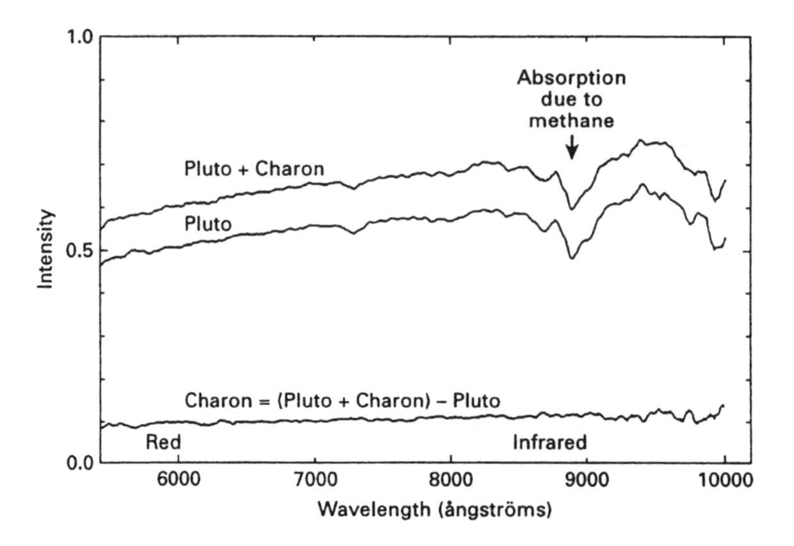

FIGURE *3.9*

The separate spectra of Pluto and Charon measured by Uwe Fink and Michael DiSanti during the occultation of Charon by Pluto on March 3, 1987: Pluto's spectrum has a strong absorption due to methane, near 8,900 Ångströms, while Charon's does not. Charon's spectrum is flat across the whole wavelength range covered, indicating that Charon's surface has little color. By contrast, Pluto's spectrum gets stronger toward the infrared end of the spectrum, indicating that it is slightly reddish. Note also that the methane-absorption feature near 8,900 Ångströms (0.89 microns) is not present in Charon's spectrum, revealing that Charon's surface contains little or no methane ice. (Adapted from U. Fink and M. A. DiSanti, *Astronomical Journal, 95,* 1988)

Buie, a long-time Plutophile with great skill and intense ambition, was incredibly disappointed at missing the March 4th event, so he secured telescope time in Mauna Kea for the upcoming April 23rd event—the very next one visible at night from Hawaii. Along with Dale Cruikshank and their colleague Larry Lebofsky, Buie tried again. This time, the winds on Mauna Kea were benign, and the team obtained a beautiful infrared spectrum, which confirmed the presence of water ice on Charon as shown in Figure 3.10.

When all of the groups had published their findings, it was clear that the Pluto–Charon pair were no more twins in terms of their reflectivity, color, and composition than they were in their size. Whether these differences were due to something that stretched back to their origins, or instead to differences in their evolution, which four billion years of time had brought to bear, wasn't clear, but the impact was: In just a few hours of observing time spread over March and April 1987, four research groups at four different telescopes had revealed that the dichotomy in size, reflectivity, color, and composition between Pluto and Charon was as striking as the dichotomy between the Earth and Luna. In the frozen half-twilight of the far outer planetary system, where conventional scientific wisdom predicted numbing uniformity, Nature proved She was not listening to conventional wisdom.

THE FACE OF PLUTO

Audacious is a good word for what came next. It's one kind of miracle to measure the size of distant, little points of light too small to actually

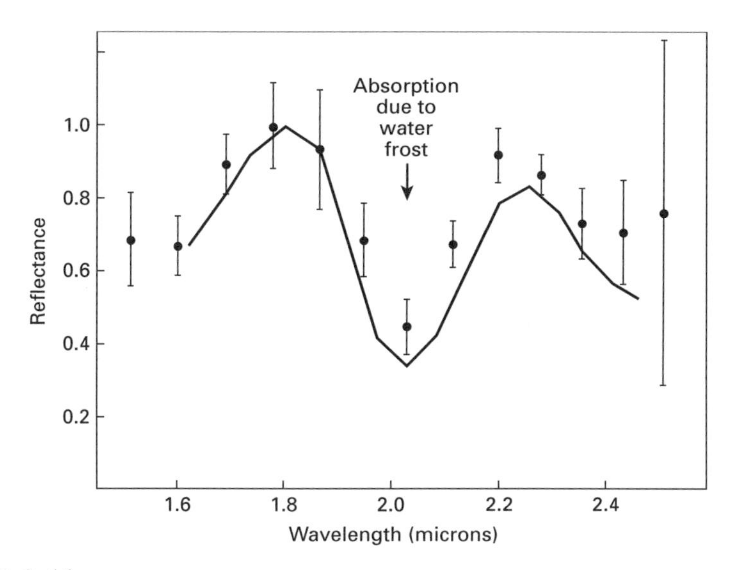

FIGURE *3.10*

The infrared spectrum of Charon, measured by Marc Buie and collaborators on April 23, 1987. This spectrum was obtained during a total eclipse of Charon, by subtracting Pluto's spectrum from a Pluto+Charon spectrum, as described in the text. The measurements for Charon are marked by black dots, with vertical error bars through them. The solid line is a laboratory spectrum of water frost. Notice the good correspondence, particularly the dip near 2 microns wavelength, which is the characteristic signature of this type of frost. (The vertical reflectance scale has been set arbitrarily to read 1.0 for the measurement at 1.8 microns.) (Adapted from Buie et al., *Nature, 329,* 1987)

resolve, and it was an even more impressive miracle to uncouple their blended spectra to see the distinct character of each one, apart from its close-lying companion. But this! Now some of the mutual event contingent planned to attack the mapping of Pluto, that grain of sand at 1 kilometer (1,100 yards), using the disk of Charon as their scimitar.

To be fair, this was not the first attempt to learn something about the distribution of surface markings on Pluto. Back in the early 1970s, John Fix had tried to do this from his lightcurves, and in the early 1980s, Bob Marcialis undertook a similar project as his master's thesis. Marcialis's project got the better of the two results.

What Marcialis did was to look at Pluto's lightcurve and ask, "What is the simplest distribution of surface markings that would be consistent with Pluto's rotational lightcurve?" This is just the kind of approach astronomers love. It's spare, it's economical, and if Nature holds high the same attributes, then it might even be a good approximation of reality.

When Marcialis examined the lightcurves that Hardie and Walker, and later Andersson and Fix had obtained, he saw two prominent dips, one on either side of the lightcurve peak. Marcialis saw that this could be simply explained by two large (but different-sized), circular dark spots on the surface. This wasn't the only way the lightcurve structure could be explained, but it did fit his data. In keeping with the long-standard assumption that the changes in the lightcurve's amplitude and in its average value were both due to the increasing amount of equatorial terrain that could be seen on Pluto as it moved around the Sun, Marcialis surmised that the dark spots were near Pluto's equator.

To explore this further, Marcialis coded up a computer program to compare the various lightcurves with those that would be produced by various "Plutos" with various bands and spots on them. The trick was to see what combination of spots would produce a lightcurve that best mimicked the actual data. In his computer, Marcialis varied the locations, sizes, and brightnesses of the spots and bands, always searching for solutions that fit the data using the fewest spots. Marcialis realized that if the brightnesses, extents, or locations of the surface features had changed with time, then almost any combination of spots was possible. Because there was no compelling evidence for the time variability of Pluto's surface, however, Marcialis assumed that Pluto's surface markings were essentially unchanging over the 30-year timespan during which lightcurve data had been obtained. Marcialis's resort to a Pluto that presented a constant surface appearance from the 1950s to the 1980s is an example of a kind of scientific reasoning called "Occam's razor," which is common in research. In effect, the razor states that given two or more plausible ways to solve a problem, the approach that makes the fewest assumptions is preferred.

Marcialis's first results were presented at technical talks in 1982 and in his 1983 thesis report to Vanderbilt University. Marcialis discovered that the decrease in Pluto's average brightness (observed over the previous few decades) almost certainly meant that the polar regions must be brighter than the equator, which his computer model wrapped in a darkened band. This model was reminiscent of Walker and Hardie's finding that Pluto's polar regions were brighter than its equator. Superimposed on the dark equatorial band, Marcialis's best-fitting map displayed darker spots that depressed the lightcurve downward from its peak, as shown in Figure 3.11.

Marcialis's use of rotational lightcurves provided some exciting and tantalizing results, but with the onset of the mutual events, it became possible to obtain even better information about the shapes, brightnesses, and locations of Pluto's surface markings. The new technique (see Figure 3.12) was to use the passing of Charon's disk in front of various regions to measure how brightness varied from place to place across Pluto's Charon-facing hemisphere.

This kind of modeling is done by placing tiles of varying brightness

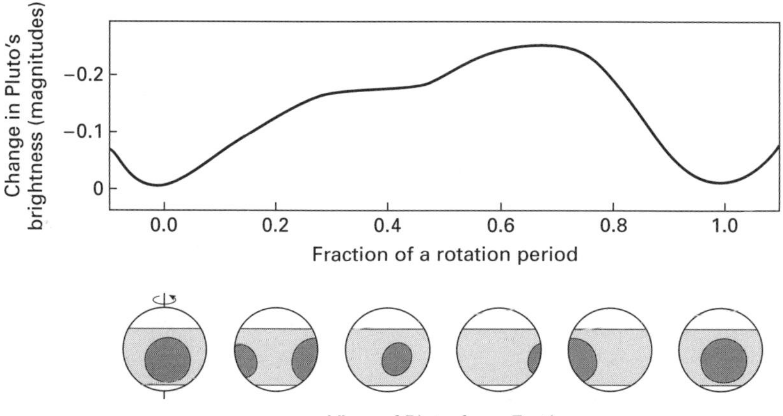

FIGURE 3.11

The uneven variation in Pluto's brightness as it rotates on its axis can be explained by the presence of areas on the surface that are darker or lighter than average. (Adapted from R. L. Marcialis, *Astronomical Journal, 95,* 1988)

on a computer-generated Pluto, and then blending those idealized surface tiles into a realistic pattern that reproduces the mutual-event lightcurves.

Like a lot of things we have run across in this chapter, the devil is in the details. For example, once again, there are those pesky shadows to keep track of. Then there is the fact that Pluto rotates on its axis a few percent during the event, causing a bit of change to its brightness. There is also the fact that Pluto is a rounded sphere, rather than the two-dimensional disk it appears to be when projected on the sky. As a result, shadows are curves, and spots and other features can appear to be foreshortened near the limb. And what if Charon has markings too? And isn't Charon rotating? Won't these factors affect the results? Yes, so to be sure of accurate results, the appearance of Charon has to also be modeled. And, just to show the real knottiness of the game, the exact radii of Pluto and Charon must be known, but these are derived from the same data, so, in effect, the model has to solve for the surface markings and the radii simultaneously. The list of complicating factors that have to be taken into account by such a computer model goes on and on, but we won't.

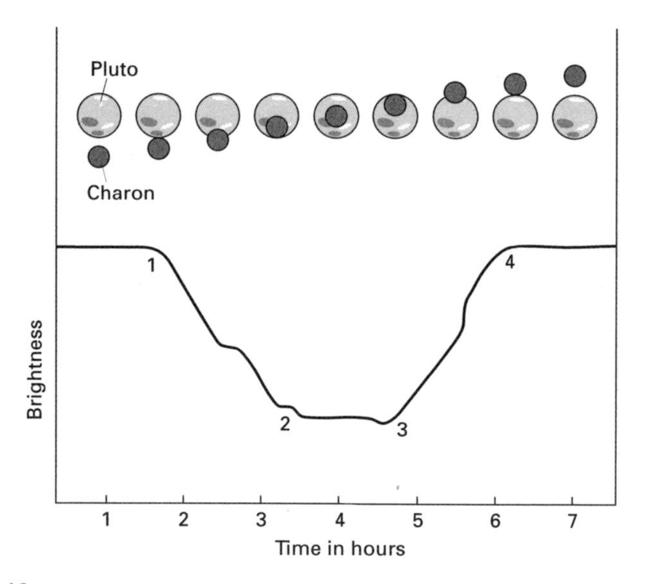

FIGURE 3.12

The presence of lighter and darker areas on Pluto's surface cause small bumps in the lightcurve recorded during the mutual events.

No one said it would be easy to map that little rotating grain of sand a kilometer away by using another rotating grain of sand, about half its size, passing in front of it every 6.4 days. Still, if the programming could be done, what a prize these maps would be!

The researchers who took on this work literally put thousands of hours into the details of computer coding, an almost-rabbinical approach to reduction of data from the mutual events, and careful analyses of the uncertainties in their results.

Marc Buie and Dave Tholen (Figure 3.13), along with Keith Horne, all fresh postdocs at the time, undertook one of the most careful mapping projects. MIT Ph.D. student Eliot Young (Figure 3.14) and his young Ph.D. advisor Rick Binzel undertook another. Both groups published their findings in 1992. These are shown in Figures 3.15 and 3.16.

Buie and company used a combination of mutual-event lightcurves and rotational lightcurves stretching back to 1954 as inputs to their model. Young and Binzel relied solely on mutual-event lightcurves, which limited their results to the Charon-facing hemisphere of Pluto. Why this limitation? Because Charon's orbital period is known to be identical to Pluto's rotation period, Charon hovers over one spot on Pluto. As a result, all of the eclipses of Pluto by Charon were over the hemisphere centered on that place. In contrast, Buie's team used both rotational lightcurves and mutual-event lightcurves, which allowed them to map both hemispheres of Pluto.

The two groups applied different mathematical algorithms to determine the distribution of markings on the surface, and they used different data sets as inputs to their models. When all was said and done, the two maps that were obtained contained both similarities and contradictions.

Despite their differences, the maps produced by these teams (and the maps produced by several other groups that tried their hand at this) were a remarkable triumph. Through careful observing, painstaking data reduction, and intensive computer modeling, the maps revealed a number of interesting attributes, many of which the Hubble Space Telescope later showed were in fact real.

Each map showed that reflectivities on the surface varied greatly, ranging from a low of about 15% (rather whitish dirt) to more than 70% (like a bright snow). Each group found evidence for an extensive, bright southern polar cap, with a kind of central dark notch. Both also found an almost featureless darker area between the southern cap and the equator, as well as a somewhat brighter region with interesting lon-

FIGURE *3.13*

Marc Buie (left) and David Tholen (ca. 1983): Buie and Tholen have been intensively involved in Pluto studies since the early 1980s, and they took a major role in the observation and analysis of the mutual events. (Courtesy Marc Buie)

FIGURE *3.14*

Eliot Young, as an MIT graduate student, worked with his Ph.D. professor, Richard ("Rick") Binzel, on one of the prominent projects to produce a surface map of Pluto based on lightcurve data from the mutual events. (Courtesy Seth Shostak)

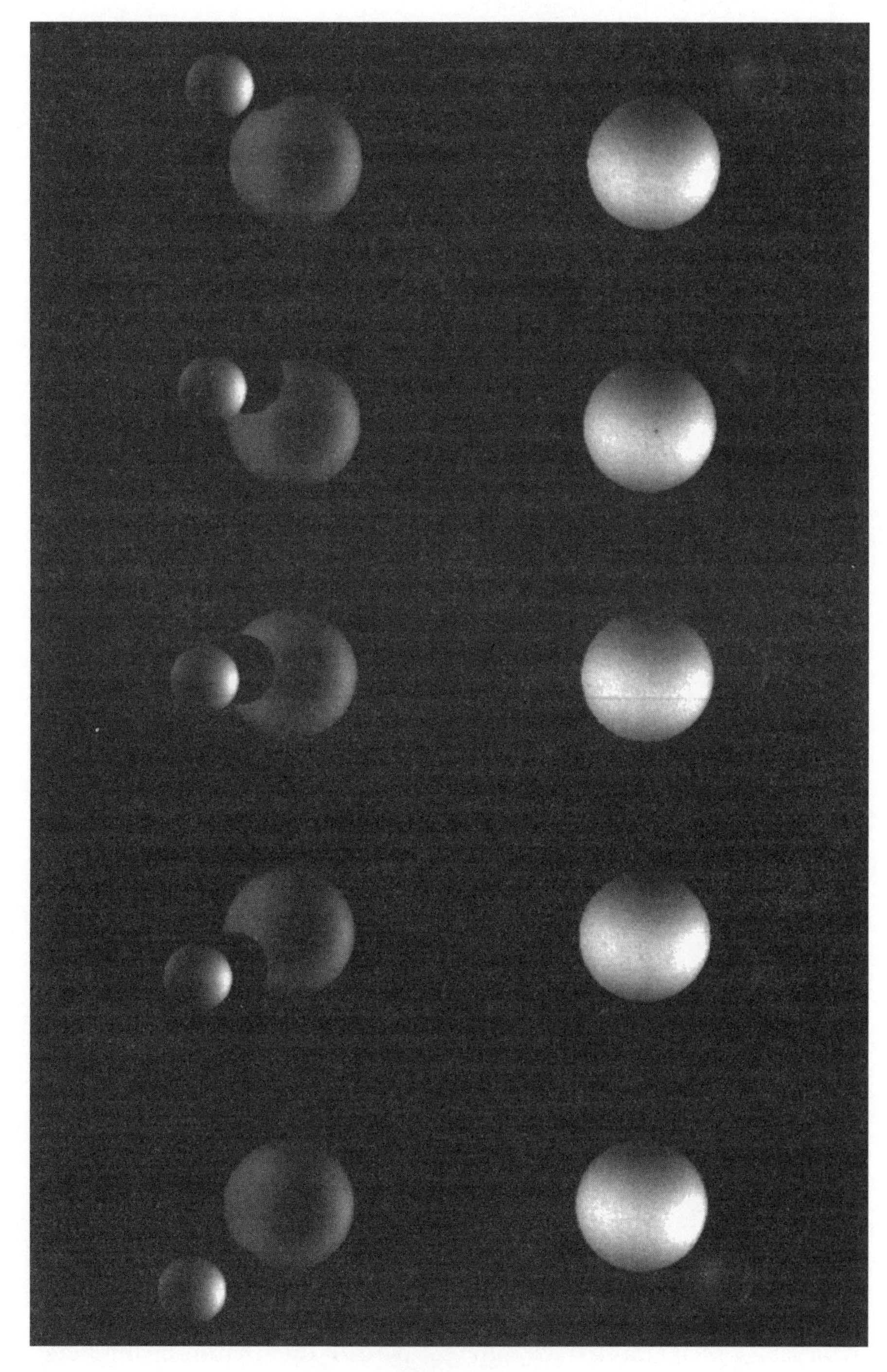

FIGURE 3.15

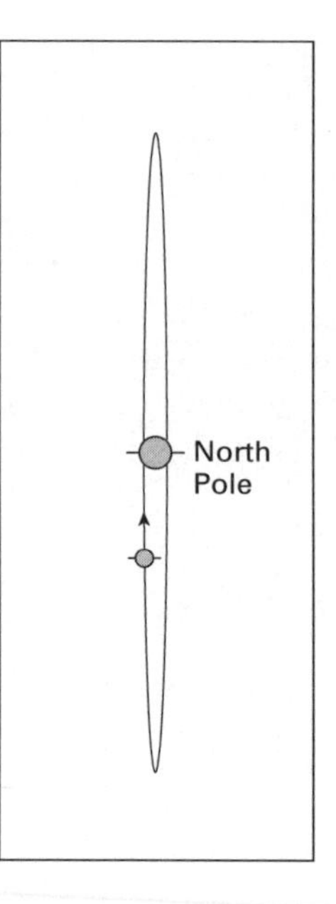

FIGURE 3.15 *(continued)*

Two computer-generated simulations and a diagram of mutual events between Pluto and Charon in 1988: The simulations—made by Keith Horne, Marc Buie, and David Tholen—resulted from the analysis of lightcurves from previous mutual events. Horne and his colleagues used these to construct surface-feature maps that would produce the observed lightcurves. The left-hand series of images runs from top to bottom and shows a transit of Charon across Pluto on February 6, 1988. The images on the right run from bottom to top, following the motion of Charon as it is eclipsed by Pluto on February 9, 1988. North on the sky is toward the top, and Pluto's north pole is tipped over 90 degrees toward the right, as shown in the diagram. (As shown to a meeting of the American Astronomical Society, January 1989)

gitudinal structure at low northern latitudes. Although the details of these regions differed in the two maps, it's fair to say that equatorward of about 45 degrees north, the two groups reached similar findings about the gross distribution of bright and dark regions. Poleward of 45 degrees north, however, there was major disagreement. Young and Binzel's modeling predicted a tiny but bright northern cap. Buie, Tholen, and Horne's computer model predicted a less bright, but more extensive, northern cap.

Which map was correct? Which was wrong? Both groups agreed it was not that kind of contest. For one thing, each group used different data sets, which emphasized different parts of the Charon-facing hemisphere. Further, because each model involved more than 20 variables

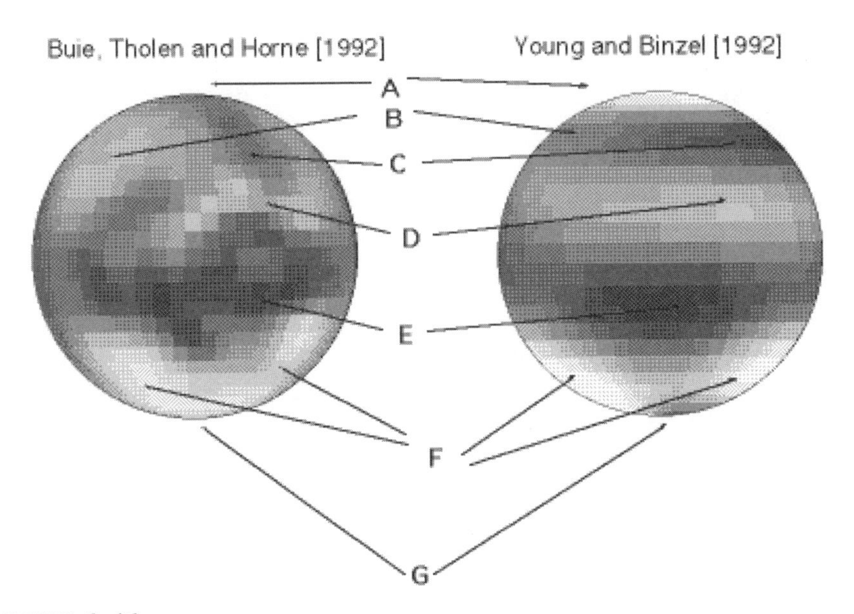

FIGURE *3.16*

Two different attempts to construct surface maps of Pluto's sub-Charon hemisphere from lightcurve data: The two are based on different techniques and independent data sets. Young and Binzel utilized data taken during transits of Charon across Pluto during the mutual events. Buie, Tholen, and Horne used both mutual-event data and information from Pluto's rotational lightcurve. Though there are clearly some differences (for example the area labeled B), there are nevertheless some striking similarities, including a large, bright, south polar cap (F). (Courtesy Eliot Young)

that had to be simultaneously optimized, each group recognized that neither result represented a perfect solution. Still, despite their uncertainties, both the value and the technical tour de force of the mutual-event maps were unquestioned: The maps revealed a planet that seemed to have at least one and possibly two polar caps, various dark bands and bright spots, and some of the starkest combinations of bright and dark terrain in the entire solar system.

CHAPTER

4

ANOTHER SNOW

If we knew what we were doing, it wouldn't be research.

ANON

"Why is Pluto bright?" It's a simple question, which, in hindsight, has a simple answer. Until 1988, however, when a scientific paper with this rather unscientific title appeared, no one had ever made the point that Pluto's snow-white surface reflectivity ultimately implies that it has an atmosphere.

That is not to say that an atmosphere around Pluto had not previously been suggested. The archives of planetary science were littered with papers containing such discussions. In fact, so many suggestions had been made, based more on speculation than on fact, that a real "prove-it" atmosphere (no pun intended) had developed about whether Pluto has an atmosphere.

The stream of papers relating to a possible atmosphere on Pluto began in the 1940s, with a landmark work by the legendary planetary astronomer, Gerard P. Kuiper. During the war years when he carried out this work, Dutch-expatriate Kuiper was the director of McDonald Observatory and about the only full-time planetary scientist working in the United States at that time. Kuiper got the idea to undertake a spectroscopic survey of all of the major satellites of the planets, in order to determine which, if any, had atmospheres. In those distant, dimly lit days before even the rudiments of modern planetary science were known, Kuiper's project was ambitious in both its scope and its vision.

An atmosphere around a satellite? How intriguing. How bold. How novel! . . . . How's that? Could satellites even have atmospheres? Well, why not? A planet's atmosphere can be generated one of three ways: by capturing some gas orbiting the Sun when the planet was formed, by outgassing or volcanic activity arising from the interior of the planet, or by vapors given off from the surface. Each of these processes could just as easily occur on a satellite as on a planet, so there didn't seem to be any reason why a satellite *couldn't* generate an atmosphere. However, there was the important detail of whether a satellite could retain its atmosphere against escape to space. If the atmosphere could escape, then it would have to be regenerated, again and again—and eventually, the supply of gas on or in the planet could run out. This is exactly what would happen to a block of ice in Earth orbit—and what does in fact happen to comets when they are active.

For an atmosphere around a planet or satellite to be stable over the eons, a planet's or satellite's gravity has to be strong enough to prevent much escape. Lots of gravity requires lots of mass, so Jupiter, with more than 300 times the mass of the Earth, is very good at holding an atmosphere. But Jupiter isn't alone in retaining an atmosphere. Similarly, the other giant planets are also good at retaining atmospheres, which is in fact part of the reason why they are gas giants. Earth, Venus, and Mars, though 300 to 3,000 times less massive than Jupiter, have enough mass to retain atmospheres, as well.

When Kuiper began his search for atmospheres around planetary satellites, all of the planets except Mercury and Pluto possessed detected atmospheres. As for the planetary satellites, however, none seemed to display obvious clouds or other signs of meteorology. In addition, the only satellite one could really inspect closely at that time, Luna, was clearly devoid of any substantial atmosphere. If Luna had ever had a substantial atmosphere, it had long since escaped to space.

What about the other satellites though? Just because no one had seen clouds and storms didn't mean there weren't atmospheres—after all, our instruments might have missed these signs from so far away, or the atmospheres might be clear and not have clouds. So, Kuiper asked himself whether it was physically plausible that some of the satellites of the outer planets might have atmospheres. Even in the early 1940s, several of these moons were known to be more massive than Luna; Saturn's Titan was even then thought to be larger than Mercury. After some simple calculations (the kind that are usually best when you have little real information to go on), Kuiper concluded that it was possible that the large satellites of the giant planets could gravitationally retain atmospheres, particularly because they were cold. Cold helps because the atoms and molecules in cold atmospheres have less energy than those in warm atmospheres, which makes escape more difficult.

Gerard Kuiper was one of those pioneering planetary observers who didn't need much motivation to simply go out and look for something new. If atmospheres might be stable around the cold satellites of the outer solar system, then he would have a look to see whether these atmospheres actually existed. Given that there wasn't any evidence for outgassing from the interiors of these satellites, or any solid evidence for ices on their surface, this was a rather speculative undertaking, but Kuiper had the resources to carry out a search, and he decided the potential rewards were well worth the effort.

Kuiper correctly surmised that an excellent way to look for atmospheres in the cold outer solar system was to search for spectroscopic absorptions due to methane. The two rationales for looking for methane, in particular, are that it is a cosmically abundant molecule, and that it remains a gas even at the very low temperatures prevailing in the outer solar system. In addition—best of all—even a little methane creates a huge absorption feature in a planet or satellite's spectrum, so it can be detected even in trace amounts. If the satellites had atmospheres, even thin ones, Kuiper bet, they would probably show methane-absorption signatures.

So, in the winter of 1943–1944, Kuiper used the large, 82-inch reflector telescope at McDonald Observatory to search for methane absorptions in the spectra of the ten largest planetary satellites, and Pluto (which he included almost as an afterthought). Because PMTs and CCDs weren't available then, astronomical spectra were recorded on photographic plates. It was the Dark Ages, though it surely didn't seem so at the time.

By the time Kuiper's project was complete, he had found three interesting things. First, Saturn's largest satellite, Titan, very clearly registered a methane signature, indicating for the first time that an atmosphere had been found around a moon! Second, Kuiper also found that Titan was something of an anomaly. None of the other satellites that he had studied showed a clear methane signature. Apparently, atmospheres of satellites were either rare, or at least below his detection limits, and he correctly surmised that the ability of gases to easily escape from satellites was probably an important factor in the lack of atmospheres around most. Finally, Kuiper discovered that his equipment wasn't sensitive enough to explore the methane question with regard to two of the faintest targets he looked at: Neptune's moon Triton, and

Pluto. Being conservative, Kuiper didn't say Triton and Pluto didn't have detectable methane absorptions. His data just weren't good enough to definitively conclude this. So what Kuiper did was to place neither a "yes" nor a "no" in the atmosphere column for these two distant bodies. For each, he just entered a "?" in his census of atmospheres.

THE LOW ROAD

Decades passed between Kuiper's paper and Dale Cruikshank's dramatic and unequivocal discovery of Pluto's surface methane ice in 1976. (Cruikshank, we note, was one of Kuiper's later students.) Nonetheless, Kuiper's findings raised the question of whether Cruikshank and company's discovery of methane ice really meant that Pluto has an atmosphere.

The situation at Pluto wasn't as clear as the situation at Titan, where the surface temperature was above the freezing point of methane, and methane gas could thus easily sublimate from methane ice to form an atmosphere. By contrast, Pluto is so far from the Sun, and therefore so weakly warmed by its light, that its surface temperature is below the freezing point of methane. As a result, the methane absorptions Cruikshank detected would largely derive from frozen ice, rather than in gaseous form. The operative questions, then, were how much, if any, of the methane would sublimate to gas, and how much of Cruikshank's absorption signature was due to methane gas. This wasn't an easy question to answer.

For many substances, a scientist with such a question could just enter into a computer the published laboratory gas and solid spectra and use these to find out how much gas and how much solid ice he or she needed to match the observations. (This technique, in its simplest form, is akin to mixing paints of primary colors to match a given hue.) The problem with Pluto's methane, however, was that the library of such low-temperature methane spectra wasn't complete. (After all, why would *anyone* have thought to measure the spectrum of methane ice and gas just above absolute zero?) Worse, from the small amount of data that did exist back in the late 1970s, methane spectra didn't seem to reveal clear distinctions between a solid and a gas unless it was observed at very high spectral resolution. Thus, the goal became to get an infrared spectrum of Pluto at high spectral resolution. This is easily said, but the telescope and detector technology to do this wasn't available in the 1970s, or the 1980s. In fact, it was 1994 before this was accomplished.

So what did those ardent astronomical explorers of the wild black yonder of the 1970s and 1980s do, in the meantime? They did what astronomers commonly do: They resorted to more clever means to attack the question.

First, theorists attacked the problem by simply calculating how much gas methane ice would generate at Pluto's temperature. To make this calculation, they simply had to look up something exotically called the "vapor-pressure relation" for methane ice, put the formula into a computer, and plug in the temperature. The result: a predicted atmospheric pressure. Unfortunately, vapor-pressure relations are exponentially sensitive (read: very, very, *very* sensitive) to the assumed temperature of the ice. In fact, in the temperature range expected for Pluto, the methane-vapor pressure doubles every time the temperature of the ice increases by 2 degrees Kelvin.[12] So to get a meaningful result, it was necessary to first have an accurate surface temperature for the methane ices on Pluto.

And what *is* the surface temperature of Pluto? Well, because the technology wasn't available in the late 1970s and early 1980s to measure the feeble heat emitted from Pluto across the 5 billion kilometers that separated it from Earth, Pluto's temperature also had to be calculated. That depended on knowing how well the surface reflects light and emits heat, which were also not known at the time. So the theorists made their best guesses, started calculating, and concluded that the surface was probably between about 40 degrees Kelvin (–390 F) and 65 degrees Kelvin (–340 F). But whether the answer was 40 Kelvin or 65 Kelvin made an enormous difference—a factor of about 1000 in terms of methane vapor pressure.

Things got even worse whenever someone pointed out that the methane might not be pure. This issue posed problems that the amount of atmosphere generated by the frozen methane on Pluto's surface depends on the methane's purity. Clearly, the vapor-pressure calculations weren't good enough to use as a definitive guide as to whether or not Pluto has an atmosphere. There were just too many uncertainties in the result.

A second, independent, attack was mounted by the observers (who generally distrust theoretical results, anyway). Recall that back in

---

[12]The Kelvin temperature scale used by scientists starts at *absolute zero,* which is equivalent to –273 Celsius or –463 Fahrenheit. Using the Kelvin scale avoids negative numbers for low temperatures and provides an easy reference point to the true zero point of nature. Each degree Kelvin is the same as 1 degree Celsius, or 1.8 degrees Fahrenheit.

1970, John Fix and his collaborators had obtained that first, tentative, 21-point spectrum of Pluto revealing its red color, and a possible little upturn on the blue (i.e., left) side of the spectrum. Fix and his coworkers suggested at the time that the upturn might be due to *Rayleigh scattering*—the effect that makes the Earth's sky blue.

The operative thing about Rayleigh scattering is that it takes a whale of a lot of atmosphere to generate the blue sky above our heads every day (if it didn't, you'd see a blue roomful of air). What this means is that if Rayleigh scattering was indeed the cause of the upturn in Fix & Company's spectrum, then Pluto had to have *quite* an atmosphere. Of course, you already know from Chapter 3 what Fix and company did not: that the little blue upturn in their spectrum was an instrumental artifact. But no one knew that in the late 1970s, so as a result, a whole series of theoretical speculations were launched about this upturn.

Fortunately, before too many theorists wrote papers about the implications of Fix's blue spectral upturn, McDonald Observatory astronomers Ed Barker, and Bill and Anita Cochran, obtained a far better spectrum of Pluto in 1979, showing that Fix & Company's result was spurious. The implication that there wasn't any strong Rayleigh scattering in Pluto's atmosphere was clear: If there were an atmosphere around Pluto, it had to be pretty tenuous.

What other means might there be to get at the question of whether Pluto has an atmosphere? Well, if there had been an opportunity to fly one of the *Voyager* spacecraft to Pluto, all of this could have been easily wrapped up using its onboard instruments to study Pluto at close range. Similarly, if Pluto's atmosphere had generated variable hazes or clouds, the atmosphere could have been detected by the ever-changing, day-to-day variations that their presence would have induced on the lightcurve. However, the lightcurve showed no such variations, from which some concluded there was no atmosphere, but which others more correctly concluded meant only that if there were an atmosphere, it was very clear and lacked the kind of variable meteorology Earth and Mars display.

All these twists and turns are fun to enjoy from the easy-chair that 1990s hindsights provide, but in the 1970s and 1980s, it was a difficult and frustrating challenge. Honest people were doing their best with limited tools to glimpse the true nature of Pluto. However, as is often the case with tough research on the frontier of astronomical technique, one step forward was often followed by another step backward.

Without the capability to obtain the high-resolution infrared spectra that were needed to accurately determine how much of the methane absorptions were due to ice and how much was due to gas, things bogged down.

A CLEAN MACHINE

All this frustration was to come to an end in the summer of 1988, when a rare occultation by Pluto would definitively settle the issue of Pluto's atmosphere through yet another clever device. However, before we see what was discovered in that occultation, there is one more tale that's worth recounting.

It began with a question. It was the same question that we began this chapter with: *Why is Pluto Bright?*

By all rights, Pluto shouldn't be bright because the very methane that lies on its surface should react with ultraviolet sunlight to create more complex carbon- and hydrogen-bearing (i.e., hydrocarbon) molecules, such as ethane, ethylene, and acetylene. These more complex hydrocarbons are dark, whereas methane is bright. As a result, methane turns into a blackish–reddish residue if left unshielded from ultraviolet sunlight.

The fact that methane would react and turn to darker stuff had been firmly established in laboratory simulations of conditions in the outer solar system. In fact, studies indicated that the time for the methane on Pluto's surface to turn black was just a hundred thousand years—a veritable wink of the eye compared with the age of the solar system.

So how could Pluto's methane stay bright? There were really only three viable alternatives: (1) Pluto could have an atmosphere that shields the methane from the ultraviolet rays that would darken it; (2) Pluto could have volcanos that routinely repaint the surface with fresh, bright methane snow; or (3) Pluto could have an atmospheric mechanism for cleaning the methane.

All three of the alternatives implied an atmosphere of one sort or another, but the atmospheric-cleansing mechanism made the most sense. In part, this was because it was the simplest explanation. Also, it didn't invoke any new claims about Pluto—the way volcanos did. Instead, the atmospheric-cleansing mechanism was just a simple byproduct of Pluto's orbital cycle, which offered a natural explanation for a long-standing mystery about Pluto's changing lightcurve.

The gist of the atmospheric-cleansing scenario was this: Regardless of the present temperature on Pluto, whether it was 40 Kelvin or 65

Kelvin, there would be some vapor pressure, and that would make an atmosphere, even if it were very tenuous. After each perihelion, the temperature would start to drop as Pluto drew farther and farther from the Sun and cooled. As the temperature dropped, the exponential dependence of the methane vapor pressure on temperature would cause the amount of methane in the atmosphere to rapidly decrease. It was estimated that the atmospheric bulk might drop by factors of 100 to 1,000 within 20 years after perihelion.

And where would the atmosphere go? It would condense out on the cold surface, as a frost layer perhaps a few centimeters (1 or 2 inches) deep. Because methane frost is bright, Pluto would brighten after each perihelion as the planet receded from the Sun and cooled. Then, for about two centuries, the atmosphere would lie there, frozen as a snow on the surface, while Pluto glided out to its farthest point from the Sun, rounded its orbital bend, and then once again approached the relative warmth of its next perihelion. All the while, of course, the fresh methane snow would be subject to ultraviolet sunlight, which would begin to darken it. But, then (this is the best part), as the surface warmed again as Pluto approached its next perihelion, the methane would sublimate (i.e., it would be converted from ice to gas in the same way that a block of dry ice "smokes"). This sublimation would regenerate the atmosphere, but it would leave behind the inert byproducts of ultraviolet processing. As such, the methane snows created two centuries before would disappear from the surface. As they did, they would leave behind their darkened residue, something like the way that a snowfall darkened by road salts evaporates by solar power and leaves behind its dirty residue. This is shown in Figure 4.1.

Each Plutonian orbit, the same cycle would reoccur. As perihelion approached and the atmosphere formed, Pluto would darken, due to the removal of the overlying burden of snows that had condensed onto the surface two centuries before. If the snows weren't entirely uniform over the planet, then as they were converted back into the atmosphere by heating, the thinnest snows would likely disappear first, making Pluto look spotty as it approached perihelion. After perihelion, as the atmosphere once again cooled and condensed back onto the ground, the surface would once again brighten with the freshly fallen methane snow. In effect, the atmosphere would act as a laundry, cleansing the uppermost layers of the surface every 248-year orbit, and in doing so, keep Pluto bright. Because only a few centimeters of UV-processing residue would be created over the 4.5 billion year age of the solar system, there didn't seem to be any problem with the methane below the

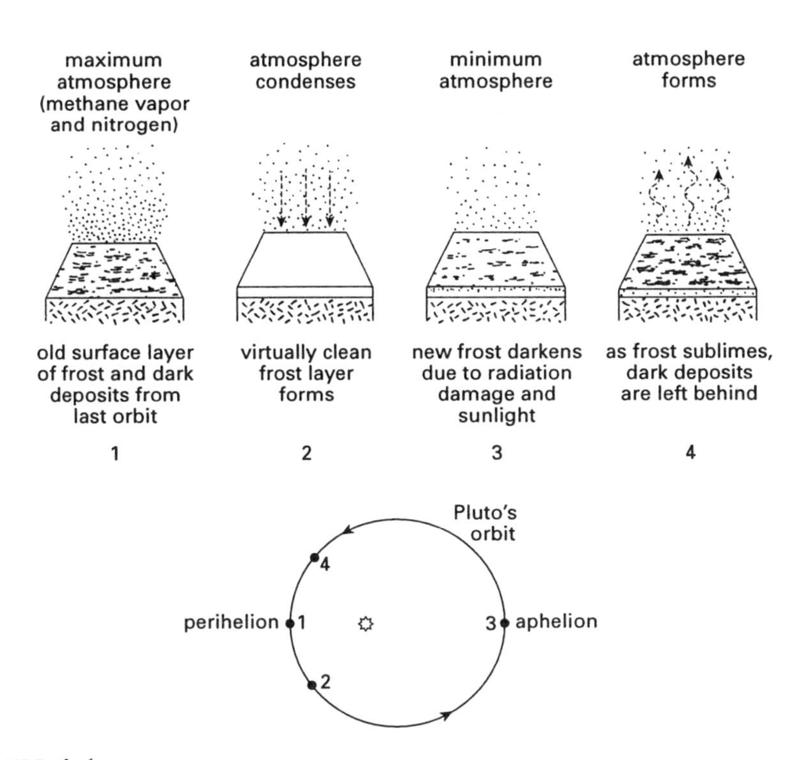

FIGURE *4.1*

The cycle of seasonal change and frost deposition affecting Pluto's surface: The top part of the diagram shows the effect on the surface of the seasonal changes. The lower part shows the approximate position of Pluto in its orbit corresponding to each stage, though it is not possible at present to predict the exact timing. (1) As Pluto reaches perihelion, increased heating by the Sun causes nitrogen and methane to sublimate from the surface into the atmosphere. The surface becomes darker as a photochemical sludge, which does not sublimate, is left behind. (2) As Pluto's distance from the Sun increases after perihelion, and its surface temperature falls, nitrogen and methane condense out of the atmosphere onto the surface, creating a fresh bright layer of new frost. (3) The methane frost on the surface darkens in color as ultraviolet radiation from the Sun causes chemical changes to take place. (4) The atmosphere starts to re-form as Pluto's orbit carries it closer to the Sun and the surface grows warmer again. The seasonal cycle repeats.

residue wafting through it when heated at perihelion. In short, if the supply of methane near Pluto's surface was fairly plentiful, the little Plutonian laundry could go right on, for eons.

What a nice, tidy story. And the thing that made it work best for people was not that the numbers behind the story worked, but that the physics were so simple and logical. Apparently, without realizing it, observers had been seeing the sublimation of the previous orbit's snows into the atmosphere, exposing the residue layer below. For it was the orbital cycle of atmospheric collapse, dormancy, and rebirth that offered a wonderfully consistent explanation for the long-perplexing mystery of why Pluto's reflectivity had been declining ever since Hardie and Walker's 1950s measurements! There was even some evidence that Pluto had grown redder as it grew darker, which matched with the atmospheric laundering idea as well.

So when someone finally asked, in 1987, "Why is Pluto bright?" the silence was almost deafening. The whole question had been overlooked during the dozen years since methane had been discovered. Once the issue was exposed, there just didn't seem to be an easier explanation for how a methane-coated Pluto resisted blackening in the Sun than the atmospheric laundry cycle.

In one fell swoop, atmospheric laundering seemed to explain a lot that hadn't fit together before, and so it captured the stage and convinced most people that we should not only expect an atmosphere on Pluto, but a dynamic one at that. The operative questions thus became how much of an atmosphere would Pluto generate, and how could it be definitively detected? The answers weren't long in coming.

THE HIGH ROAD

Although the "laundry's" logical framework convinced many planetary astronomers that Pluto's methane signature certainly did imply that Pluto had an atmosphere, planetary observers insisted that *nothing* substitutes for direct evidence. They were right, so detect it they did.

The clear and unambiguous detection of Pluto's atmosphere occurred a fine June day in mid-1988. As you might have expected, when the detection was actually made, it revealed some real surprises.

The definitive detection of Pluto's atmosphere was made by observing a stellar occultation by Pluto. This technique works because if a planet has an atmosphere, even one less than a million times as dense as Earth's, then that atmosphere will refract starlight passing through it, thereby imprinting an unmistakable signature of refraction, which in turn belies the presence of the planet's atmosphere.

Thus, a very good test for whether Pluto (or any other world) has an atmosphere, is to observe a star being occulted by the planet. A star occulted by an atmosphereless world (such as the Moon) would disappear almost instantaneously as the limb of the planet quickly cut off the light of the star. However, if the world occulting the star sports an atmosphere, refraction causes the signal to decline much more gradually. How gradually depends on the density and the thickness (which astronomers like to characterize by something called the "scale height") of the atmosphere. So in an occultation observation, the experimental objective is to obtain a running record of the brightness of the star as it passes behind the planet, and to determine whether it just winks out in a tiny fraction of a second, or instead slowly fades, perhaps over tens of seconds, or even minutes.

Stellar occultations have been used to probe planetary atmospheres for decades. As we pointed out, however, Pluto's small size and great distance conspire to make such occultations very rare. The first serious attempt to observe an occultation by Pluto occurred when Ian Halliday, Bob Hardie, and Otto Franz set out in 1965 to determine Pluto's diameter. Although this troika didn't set out to search for an atmosphere, they might just have found one. Unfortunately, the track of Pluto's shadow on the Earth due to the occultation couldn't be predicted very accurately in those days, and all Halliday and team observed was a tantalizing near-miss. As seen in their telescope, the hoped-to-be-occulted star passed about one ten-thousandth of a degree (3,300 kilometers, or 2,060 miles) off Pluto's limb. Close, but no cigar.

The next try came in April of 1980. By this time, it was clear that the occultation presented an opportunity to search for Pluto's possible atmosphere, but the event predictions indicated that most of the shadow track would lie over the south Atlantic, in daytime, with only a slight chance that observers at the south end of Africa might be able to catch the event at night. This was of course the event Alistair Walker luckily observed, but from Walker's vantage point, the starlight was extinguished by Charon, instead of Pluto. Although he didn't achieve his intended goal of observing an occultation of Pluto, he was (see Chapter 3) lucky enough to catch Charon's shadow in his telescope and to accurately estimate its diameter for the first time.

A third promising event occurred in 1985, but again the shadow track fell almost entirely over open water. In fact, only two observers who were anywhere near the shadow track realized how important such an observation was to studies of Pluto: Noah Brosch and Hiam Mendelson, at the Wise Observatory in northern Israel. As luck would have it, the great Israeli–Jordanian air war was just then taking place overhead, replete with artillery fire, air-to-air rockets, cannon shots, and tracer rounds. Almost unbelievably, Brosch and Mendelson managed to observe the event despite the circumstances overhead. But when they inspected the data they obtained, they saw that the faint signal from Pluto and the target star was marred by bright flashes from the air battle taking place overhead. Their dataset was (and still is) probably the only astronomical observation ever made through a sky filled with dog-fighting Mig-23 Badgers, F-4 Phantoms, and F-15 Hornets. And although Brosch and Mendelson believed they could see telltale evidence for refraction in the way the star disappeared, the data were so contaminated with light from air-battle artillery and tracers that others who inspected the data just couldn't say whether there was something definitive there or not. Perhaps the two Israelis had detected something indicating an atmosphere. Perhaps not. Yet another try would be required.

That effort came in 1988, when predictions indicated that another star would be occulted by Pluto at night over a wide swath of land and sea. The star, which went only by the simple name "P8," for Pluto occultation candidate number eight, was wholly inconspicuous except for its appointed date with Pluto. When it became clear that the P8 event looked like it would actually pass behind Pluto (rather than off its limb), two teams of astronomers went out in force to systematically capture as much data as possible, and truly resolve the issue of Pluto's atmosphere once and for all.

Indeed, for months before the event on June 9, 1988, planning was under way. Bob Millis and Larry Wasserman of Lowell Observatory, and Jim Elliot and Ed Dunham of MIT knew the stakes were high, and organized an observing campaign that included at least seven telescopes. The event was predicted to occur over the South Pacific, Australia, and New Zealand. Knowing that factors such as Charon's tug on Pluto's position made the precise track of the shadow impossible to predict, Millis and Wasserman arranged for a fence line of telescopes to snare their prey, as shown in Figure 4.2.

During the same months in early 1988, MIT's Elliot (see Figures 4.3 and 4.4), along with his colleague David Dunham of NASA Ames Research Center, organized a wholly separate effort from the one Millis and Wasserman spearheaded. Elliot's team petitioned for and succeeded in bringing NASA's flying Kuiper Airborne Observatory ("the KAO") down to Hawaii in order to fly it out over the best-predicted ground track, 3,000 miles to the south. (This unusual move of the KAO to Hawaii for the event was costly and testifies to the scientific value of the highly anticipated June 1988 occultation.) The KAO (see Figure 4.5) had for years been armed with a 91-centimeter (36-inch) telescope. For the P8/Pluto occultation, Elliot and company placed a high-speed (i.e., rapidly framing) CCD-camera called SNAPSHOT behind the telescope that had been specially designed for occultation studies. See Figure 4.5.

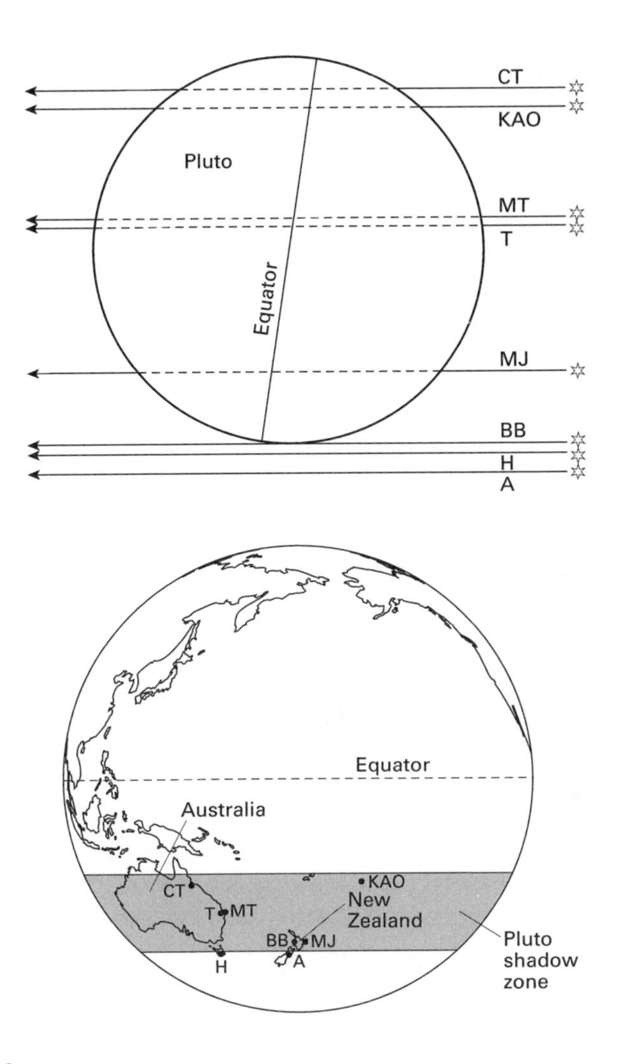

FIGURE *4.2*

Pluto's occultation of the twelfth-magnitude star P8 on June 9, 1988, was visible from a range of latitudes covering parts of Australia, New Zealand, and the South Pacific. Observations of the event were obtained from Charters Towers (CT), Toowoomba (T), Mount Tambourine (MT), and Hobart (H) in Australia; from Mount John (MJ), Black Birch (BB), and Auckland (A) observatories in New Zealand; and from NASA's Kuiper Airborne Observatory (KAO) flying over the south Pacific Ocean. The apparent track of the star behind Pluto, as seen from each of the locations, is shown in the top diagram; Pluto's shadow track on the Earth is shown on the bottom one.

FIGURE *4.3*

Robert Millis, the present-day director of Lowell Observatory: Millis and his colleague Larry Wasserman led the planning for ground-based observations when Pluto occulted P8 in June 1988. (Courtesy R. Millis)

The KAO offered Elliot some key advantages. First, it was mobile, so it offered him the chance to change his observing plan and location as the accuracy of the predicted path of Pluto's shadow improved in the days and hours leading up to the event. No ground-based telescope could do that. Second, by flying high in the stratosphere at altitudes of 40,000 feet or more, the KAO could essentially make certain that no matter what happened below, weather wouldn't shut out the whole observation effort.

Elliot and Dunham had used the KAO to capture many occultations of planets and satellites over the years. Mars, Jupiter, and Io were just a few of their trophies. Their best-known achievement, however, had come in 1977, when they discovered (quite by accident!) that Uranus has rings. Later, when Elliot was asked why he spent so much effort

FIGURE *4.4*

Jim Elliot, an experienced observer of occultations, successfully led the Kuiper Airborne Observatory (KAO) expedition to fly over the best predicted ground track for Pluto's stellar occultation in June 1988. (Courtesy J. Elliot)

FIGURE *4.5*

The Kuiper Airborne Observatory (KAO), which operated from Ames Research Center from the mid-1970s until it was retired from service in 1995: A converted military cargo plane, the KAO carried a 36-inch reflecting telescope for astronomical research. The square telescope aperture in the fuselage can be seen just in front of the wings. The KAO's successor, SOFIA, is expected to begin operations in the year 2001. (NASA)

preparing for the 1988 Pluto event, he quietly said, "We wanted to do something hard, and the only thing left was Pluto."

At the same time that Elliot and Dunham's team was preparing to use the KAO, Millis and Wasserman's team was finalizing some careful plans for the ground-based effort. The logistics alone took months to plan. Observatories had to be enticed to participate.

Mobile telescopes had to be made ready to ship to New Zealand and Australia to fill gaps in the fence line. And new observations of Pluto and P8 had to be made to refine the predicted time and exact track where the event would take place.

With Millis and Wasserman's careful, D-day-like ground-based planning, and Elliot and Dunham's successful bid to fly the KAO airplane down south to study the event, Pluto wasn't going to get away this time—and indeed, it didn't.

When June 9 came the equipment worked, the weather cooperated, and the ground-track of Pluto's shadow was close to perfection. Seven different telescopes at remote Australian and Tasmanian locations with faraway names like Mount John, Toowoomba, Hobart, and Charters Towers captured the lazy disappearance of P8 behind Pluto. Among the various ground-based telescopes that observed the event, P8's disappearance was seen to take between 70 and just over 150 seconds. In fact, the refraction effect was strong enough that some of the telescopes never even saw the signal fully bottom out, as would occur when the star fully disappeared and only Pluto remained.

The KAO saw the occultation too, rewarding Jim Elliot and his stratospheric explorers (see Figure 4.6) with 100 seconds of precious data. Aboard the KAO on the morning of the P8 event, P8's gradual fade to black behind Pluto was met with shouts of joy.

"What this means," Jim Elliot declared on seeing the data, "is that Pluto has an atmosphere!" By the next day, the news was out in papers around the world. *Sky & Telescope* reporter Kelly Beatty, who had ridden with Elliot and company aboard the KAO, wired in a story to the morning's *Boston Globe*. In a page-1 headline, Beatty and the Globe shouted, "Key Pluto Find!"

DATA MINING

The story that ran in the newspapers back in the states in early June of 1988 painted the discovery of Pluto's atmosphere as the end of a quest, but to those who observed the event, this wasn't an ending at all: It was a beginning.

FIGURE 4.6

Researchers from MIT (Massachusetts Institute of Technology) the morning after they successfully used the KAO to observe the occultation of a star by Pluto on June 9, 1988—left to right: Amanda Bosh, Stephen Slivan, Leslie Young, Edward Dunham, and James Elliot. (Photo by Ben Horita, courtesy Stephen Slivan and Leslie Young)

It was as if God had shined a searchlight through Pluto's atmosphere, and as His beam from P8 passed closer and closer to Pluto's limb, it delved deeper and deeper into the envelope of alien air surrounding this distant planet. With the data gathered during the stellar occultation, it would actually be possible to unravel the *structure* of Pluto's atmosphere. The occultation teams knew that, and they didn't waste any time.

By comparing the various data sets that had been gathered, it quickly became clear that some of the P8 lightcurves had reached down to the bottom of the atmosphere (or at least to a layer in it where the light level bottomed out). Other P8 lightcurves hadn't gotten that far because the telescopes that obtained them were located off the center track of the occultation and so had cut through the atmosphere at a more grazing angle near Pluto's poles. From the lightcurves that did reach the bottom level, it was determined that the pressure there was between 3 and perhaps 10 *microbars* (i.e., between 3 and 10 *millionths* of the air pressure at sea level here on Earth). This meant that the gas at the base of Pluto's atmosphere was very rarefied. In fact, the gas was as tenuous as the air 80 kilometers over our Earthbound heads, on the very edge of space. It was the most rarefied atmosphere ever detected by occultation. No wonder it had been so hard to detect with other, less decisive techniques! So tenuous was Pluto's atmosphere that a simple calculation showed that from the surface of Pluto, the sky overhead would probably look black.

But that wasn't all they found. Pluto's atmosphere was also revealing itself to be different from any other yet explored—and it was incredibly interesting. From the rate of the occultation lightcurve's decline (see Figure 4.7), it was possible to determine that the atmosphere extended upward many hundreds of kilometers, and that the atmospheric scale height near the surface was about 60 kilometers (38 miles). Because the pressure in an atmosphere drops by a factor of about three every time the altitude increases by a scale height, one would have to ascend about 300 kilometers (190 miles) over Pluto's surface to reach the point where 99% of the atmosphere was below. By contrast, the same point in the Earth's atmosphere would be reached at an altitude of just 40 kilometers (25 miles). In fact, atmospheric models based on the occultation results showed that traces of Pluto's atmosphere should extend many thousands of kilometers into space. Thus, whereas higher-gravity planets such as Venus, Earth, and Mars have only a razor-thin atmospheric skin (compared to each planet's size), Pluto looks more like a planet encased in an enormous spherical halo of gas.

Theorists such as Larry Trafton of McDonald Observatory and Ralph McNutt at MIT predicted that this distended atmosphere would be escaping away into the vacuum of space more easily than on any other planet. Over the 4-billion-year age of the solar system, this rate of gas loss (which comes from the sublimation of surface ices) implies that between 3 and 30 kilometers (as much as 100,000 feet) of terrain on Pluto's surface has slowly but surely been lost to space. As was becoming the norm with Pluto, this insight was something completely new.

SMOKE, OR MIRRORS?

It didn't take long before something else was discovered in the occultation lightcurves. Closer inspection of the data revealed that the slope, or rate of decline, of many of the lightcurves steepened at about the point where half of P8's light had been extinguished. What would

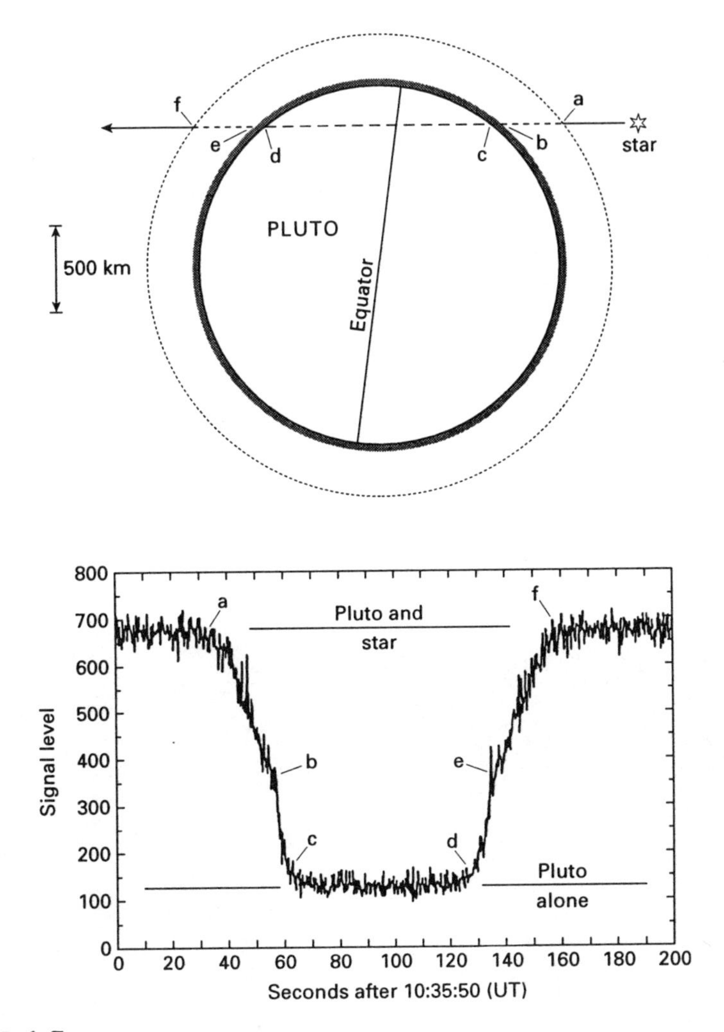

FIGURE *4.7*

The lightcurve obtained by the KAO during the occultation of a star by Pluto on June 9, 1988. The upper diagram shows the apparent track of the star behind Pluto. The starlight from P8 was not cut off abruptly, but dimmed gradually, as shown in the lightcurve below. A corresponding effect was seen as the star emerged from behind Pluto. This is a clear signature of refraction, indicating that Pluto has a tenuous atmosphere, which is represented by the outermost circle. The letters a, b, c, d, e, and f show how the progress of the lightcurve is linked to the position of the star relative to Pluto. The declining and rising parts of the lightcurve each display two slopes. When nearest the planet, the star appeared to dim and brighten more rapidly than when it was somewhat farther away. To explain this, there must be some kind of layering in Pluto's atmosphere, which could be a sharp temperature gradient, the presence of a haze layer, or both. (Adapted from J. Elliot et al., *Icarus*, 77, 1989)

cause the atmosphere to suddenly change this way? Two theories were put forward to explain this curious twist.

The first was Jim Elliot's idea. Perhaps Pluto's atmosphere contained a haze layer that extinguished some of the light. Haze layers are fairly common in planetary atmospheres. The Earth's atmosphere, for example, has many thin layers composed of aerosol droplets, ice particles, and dust high above it.[13] Working with his student Leslie Young (Figure 4.8) (Pluto-mapper Eliot Young's sister), Jim Elliot suggested that the haze would be a natural by-product of ultraviolet sunlight acting on methane in the atmosphere, just as the same ultraviolet sunlight would darken the surface methane. As shown in Figure 4.9, Elliot & Young calculated that the top of the haze layer would lie 30 or 40 kilometers over Pluto's icy terrain and would soak up about 35% of the light falling through it to the surface.

An alternative to the haze idea was hit upon separately by Von Eshleman at Stanford University and by a University of Arizona team con-

FIGURE *4.8*

Leslie Young: As a graduate student at MIT, she worked on the properties of Pluto's atmosphere. (J. Mitton)

[13]More familiar, lower-altitude examples of haze layers on Earth include fogs and pollution layers over large cities.

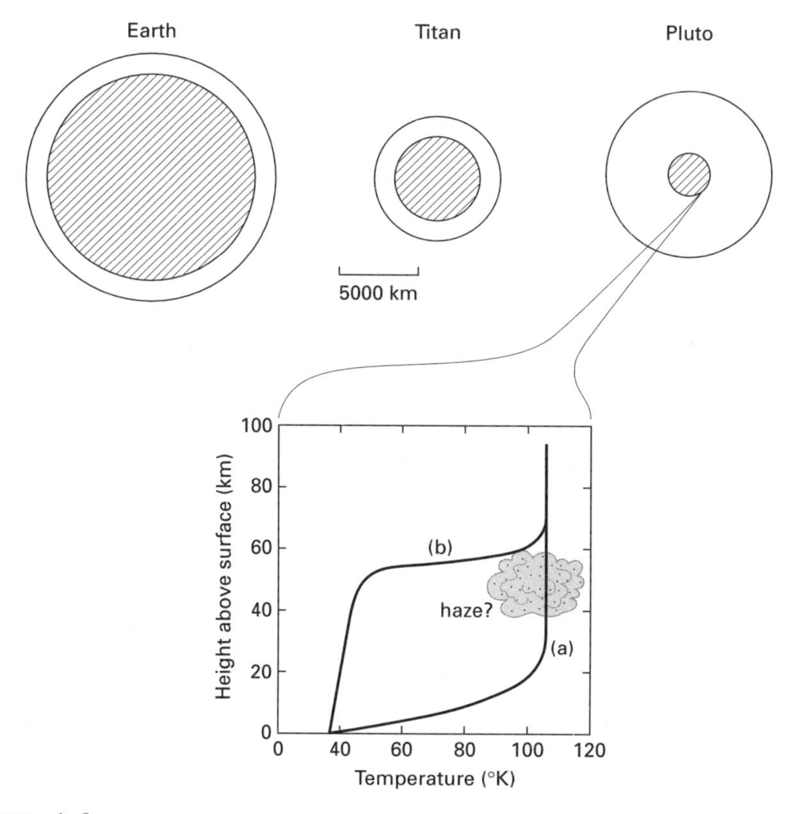

FIGURE *4.9*

Top: The extent of the atmospheres surrounding Earth, Titan, and Pluto are shown to scale for comparison. The limits shown represent in each case the upper stability limit, known as the *exobase*. Pluto's tenuous outer atmosphere is huge, compared with the small planet. (Adapted from a diagram by Roger Yelle) Bottom: Observations made during the 1988 occultation of the star P8 by Pluto probed a region of Pluto's middle atmosphere a few tens of kilometers thick (solid lines). They suggest either a layer of haze (corresponding to line a), or a steep increase in temperature with height (corresponding to line b). A combination of the two effects might also be at work. The dotted lines are extrapolations to the surface. In reality, there is also some uncertainty (about 25 km) around the zero point of the altitude scale. An average value is used for the purpose of this schematic illustration. (Adapted from a diagram by Leslie Young)

sisting of Bill Hubbard, Roger Yelle, and Jonathan Lunine (see Figure 4.10). Each recognized that the lightcurve steepening didn't *require* any haze at all. It could instead be created by an increase in temperature with height in Pluto's near-surface atmosphere. Further, the Hubbard, Yelle, and Lunine team found that this heating could naturally result from the presence of atmospheric methane (which makes a kind of

FIGURE *4.10*

Roger Yelle (left) and Jonathan Lunine (right), who like Von Eshleman, argued that the stellar-occultation data could mean that the temperature of Pluto's atmosphere increases with height above the surface. (A. Stern)

greenhouse effect). If there were enough methane in the atmosphere, and nothing to cool it, then the increase in temperature with altitude would change the amount of refraction in just the right way to create the lightcurves that had been observed by the KAO and the various ground-based groups. Yelle and Lunine calculated that if their model were correct, the upper atmospheric temperature high above Pluto's surface would reach 106 K (–270° F) or so—far warmer than the surface temperature (though still frigid by *anyone's* standards).

So, was it a haze or a thermal effect that steepened the occultation lightcurves? We still don't know. The data can be fit by either model. In fact, the two explanations aren't mutually exclusive: There could be both some haze and some temperature rise with altitude.

Interestingly, however, both the haze interpretation and the hot-atmosphere interpretation call into question one of the results obtained from the mutual events: the radius determination for Pluto. The best mutual-event estimates give radius values between about 1,145 and 1,170 kilometers (715 and 730 miles). However, if there was a haze, then Charon could have disappeared in each mutual event before it sank over Pluto's limb, and the true surface could be somewhere in the murky depths 10, or 50, or even 75 kilometers below where it seemed to be. Just as plausibly, the thermal-gradient model predicts that a kind of optical mirage could be created by the atmosphere, which could again have caused Charon to disappear from view before it actually sank below Pluto's limb. If the occultation lightcurves didn't actually get down to the hard surface, but simply to a haze deck or a miragelike level, then the atmosphere would extend farther down, and the surface pressure would be higher than a few microbars. Indeed, the pressure on the hard surface could be as great as about 50 microbars.

This issue of Pluto's true radius is of more than academic interest, at least to academics, because packing the same mass into a smaller volume implies a denser planet. A denser planet, in turn, would mean that Pluto contains even more rock, and less ice, than previously thought.

Is this actually the case? What's needed now is another occultation, so that more data can be gathered. If the event can be observed at both red and blue wavelengths, it should be possible to distinguish between a haze and a hot upper atmosphere because a haze will produce detectably different lightcurves through the two filter bands, but a temperature variation won't. The red-light data might also allow astronomers to probe deeper than the blue light allows, which should answer the question of how deep the atmosphere really is.

Of course, stellar occultations by Pluto are still incredibly rare events, and the next good candidate event is not expected until 1999. Even if that one fails to materialize, or is rained out (the KAO has been retired), it is expected that the planned NASA spacecraft to Pluto (see Chapter 7) will settle the question a decade or so later. How? A few simple pictures could tell whether there is a haze layer, how high it occurs, and how thick it is; also, onboard spectrometers could watch the Sun set over Pluto's horizon to probe the atmospheric structure without any ambiguity at all, revealing the true pressure, temperature, and haze structure right down to the surface.

SOMETHING SO FAMILIAR

One of the most interesting things about the atmospheric models developed from the stellar-occultation data was that they suggested some unexpected problems with our understanding of something very fundamental about Pluto—its surface temperature. Any freshman physics student with a hand calculator could figure out the surface temperature *expected* on a Pluto being warmed by sunlight. The formula for this is called the "radiative equilibrium temperature relation," and it isn't complicated. All you have to know is the *surface reflectivity* (that is, how much sunlight is reflected away and is therefore not available to heat the surface), the efficiency at which the surface emits heat (usually assumed to be a perfect 100%), Pluto's distance from the Sun, the Sun's brightness, and Pluto's atmospheric pressure. The mutual events had generously contributed the reflectivity. Pluto's distance to the Sun and the Sun's luminosity easily looked up in a book. The surface pressure, which the occultation results provided, is needed because it tells you how fast the ice on the surface is sublimating, thereby robbing some of the energy that would otherwise go into heating the surface. An everyday consequence like this sublimation cooling effect is the chill that comes from alcohol-based products (like an aftershave lotion) evaporating on your skin. Loading all these factors into the calculator, astronomers predicted a temperature between 50 and 60 Kelvins (–370° F and –350° F) on Pluto's surface.

Comfortingly, when the European Space Agency and NASA launched an infrared observatory called the Infrared Astronomical Satellite (IRAS) into space in 1983, it obtained thermal measurements of Pluto that were interpreted according to a hot shot young space scientist named Mark Sykes of the University of Arizona (see Figure 4.11), to give a temperature of 55 to 60 degrees Kelvin.[14]

So what's the problem? The rub came from a mismatch between the pressure measured from the stellar occultation and the pressure that would result from sublimating methane at 55 or 60 degrees Kelvin (–350° to –370°F). The occultation data implied that even if the atmosphere was about as deep as it could possibly be, the surface pressure couldn't exceed about 50 microbars. But if the surface temperature of Pluto's ices were 55 or 60 Kelvin, then the predicted pressure resulting from sublimating methane would be 100 times higher than that. The pressure measured from the stellar-occultation lightcurve and the surface-temperature information weren't compatible.

Something had to give, and the first hint of what that might be came in 1986, from an unexpected quarter. For it was then that a team of Europeans headed by Wilhelm Altenhoff of the Max Planck Institute for Radio Astronomy in Bonn obtained a different kind of temperature measurement of Pluto. This measurement was obtained using the spanking-new, high-tech IRAM (Institute de Radio Astronomie Mil-

---

[14]Notice that Sykes, like most everyone who made a significant contribution to Pluto research, including Clyde Tombaugh, was at the start of his scientific career. Pluto is largely a young person's planet!

FIGURE *4.11*

Mark Sykes, another of the young Plutophiles: Using data from the Infrared Astronomical Satellite (IRAS), Sykes makes one of the first reliable estimates of Pluto's surface temperature. (Courtesy J. Mitton)

limetrique) radio telescope perched atop the Sierra Nevada range in the south of Spain (see Figure 4.12). Altenhoff and his crew found something very different from what the IRAS data had indicated. *Their* data indicated that Pluto's surface was probably a lot colder than 60 Kelvin—more like 40 Kelvin (–390° F). But it's a dry cold.

Hardly anyone in the U.S. planetary research community knew about or appreciated the import of Altenhoff's result. Why? For one thing, as a good Euro-citizen, Altenhoff published his findings in a European astrophysics journal that few Plutophiles (all American) read. As a result, few Pluto researchers even knew of Altenhoff's finding. Further, of those who did, essentially all discounted it because the results were at odds with the then-common viewpoint that the temperature on Pluto must be near 60 Kelvin—the temperature that a methane surface would obtain at Pluto's distance from the Sun. Further, at that

FIGURE *4.12*

The 30-meter-diameter radio telescope of the Institute de Radio Astronomie Millimetrique (IRAM), located atop the Sierra Nevada mountains near Granada, Spain: IRAM is operated jointly by France, Germany, and Spain. The telescope weighs more than 800 tons. The building beside the telescope houses the control center, offices, and a dormitory where astronomers stay when working at the telescope. Using this telescope in 1988, a team of European astronomers headed by Wilhelm Altenhoff obtained the first evidence that much of Pluto's surface is colder than IRAS had indicated. (Ron Allen, STScI)

time the 1988 occultation of P8 hadn't occurred, so there was no atmospheric-pressure measurement to give pause to the warmer, IRAS-derived temperature. As a result, Altenhoff's prescient hint of something awry with our conception of Pluto lied fallow in the technical literature.

Meanwhile, theorists like Roger Yelle and Jon Lunine were pointing out something equally important that eventually came to bear on this puzzle. If their warm upper-atmosphere interpretation of the occultation results was correct, then it implied that something other than methane had to inhabit Pluto's atmosphere in order to fit the measured 60-kilometer (38-mile) scale height. Because methane gas would produce a much larger scale height than the data apparently allowed, the suspected second constituent had to be heavier than methane, and at least as abundant. What could it be?

There weren't many options, because most gases would be frozen solid when they reached Pluto's temperature range. About the only molecules that had the right properties were nitrogen, carbon monoxide (that poison), and oxygen. (Technically speaking, rare gases such as argon and neon would also do, but they are so scarce in the universe that it was doubtful there could be much of them on Pluto.)

A second line of evidence came to the fore in 1992–1993, when a U.S.–French group of planetary astronomers set out to check the strange cold-Pluto result obtained by Altenhoff's team. Seeing the discrepancy between the implications of the occultation data and the IRAS temperature that implied a far higher atmospheric pressure, this team obtained measurements of Pluto's thermal emission using both the same telescope in Spain that Altenhoff had used (IRAM), and another radio telescope on Mauna Kea in Hawaii. They found that Altenhoff & Company were dead right—Pluto's radio temperature was *very* close to 40 Kelvin.

When the U.S.–French team gave technical talks, and later published their findings, they pointed out that the radio results could square with both the occultation-derived pressure and the radiative equilibrium calculation of temperature *if* nitrogen or carbon monoxide, rather than methane, dominated Pluto's atmosphere. Why? Because the sublimation of nitrogen or carbon-monoxide ice at roughly 40 Kelvin would produce an atmospheric pressure right smack in the range that the occultation data demanded. Thus, if nitrogen or carbon monoxide were indeed in Pluto's atmosphere as a second major gas, it would explain why the occultation scale height was lower than predicted for methane *and* give a surface pressure in accord with both the occultation measurements and the radio temperatures. It was time to make a careful search to see whether these molecules were present in Pluto's spectrum but had been missed.

Just such a search was proposed in 1992 by Toby Owen of the University of Hawaii, and Leslie Young and Jim Elliot of MIT. Other members of their team included Dale Cruikshank, French astronomers Catherine de Bergh and Bernard Schmitt, Hawaii's Tom Geballe, and Cruikshank's former student Bob Brown, who was then working at JPL.

This wasn't going to be an easy observation. Why? Recall that we called methane a bull in a china cabinet because even a trace of it is easy to detect. By comparison, nitrogen is a church mouse—it hardly makes itself visible even when it's highly abundant—and carbon monoxide isn't much easier to detect than nitrogen. To search for their stealthy prey, the team took advantage of a new, far more sensitive infrared spectrometer that had just become available on the UK Infrared Telescope (UKIRT) at Mauna Kea. The new spectrometer, called "CGAS," was 16,000 times more powerful than the simple photometers Dale Cruikshank, Carl Pilcher, and David Morrison had used back in 1976 to find the methane. It's a good thing, too: Owen and his coworkers were going to need all the grasp CGAS could provide.

So, in May of 1993, Owen, Young, and their collaborators ascended Mauna Kea and pointed the UKIRT telescope toward Pluto for a four-night run. With CGAS's incredible capabilities, it didn't take long. Within days, the team knew they had bagged their prey because the subtle signatures of both frozen nitrogen and frozen carbon monoxide were found in their infrared spectra of Pluto.

Several months of hard work followed Owen and team's initial detection—analyzing the data, and then carefully fitting it to laboratory spectra of frozen nitrogen, methane, and carbon monoxide. Using a computer model to vary the concentrations of each of these ices, Owen (see Figure 4.13) and his team were able to determine how much of each constituent was needed to best fit their spectra. They found that nitrogen ice composed about 99.5% of the mix. In contrast, the surface contained only about one carbon monoxide molecule for every 200 nitrogens, and about one methane molecule for every 1,000 nitrogens. All the while since 1976, methane's large signature had lulled astronomers into thinking methane was the dominant ice on Pluto's surface, when in fact it wasn't.

Next, using vapor-pressure calculations that relate the composition of an atmosphere to the amount and temperatures of various ices on the surface, Owen's team transformed the surface-composition measurements (see Figure 4.14) into predictions of Pluto's atmospheric composition. They found that Pluto's atmosphere had to consist primarily of nitrogen, with minor amounts of methane and carbon monoxide.

Leslie Young added an important observational confirmation to this in 1994 when she, Jim Elliot, and several colleagues obtained the very-high-resolution spectrum of Pluto's methane absorptions that researchers had longed for for 20 years. With that high-resolution spectrum, Young and her team directly detected methane gas in Pluto's *atmosphere*.[15] For the first time, methane gas had been unambiguously

[15]Recall that the previously available, lower-resolution spectra of Pluto's methane bands couldn't distinguish the gas from the ice.

FIGURE *4.13*

With his colleagues, Tobias ("Toby") Owen discovered that the dominant ice on Pluto is frozen nitrogen. (J. Mitton)

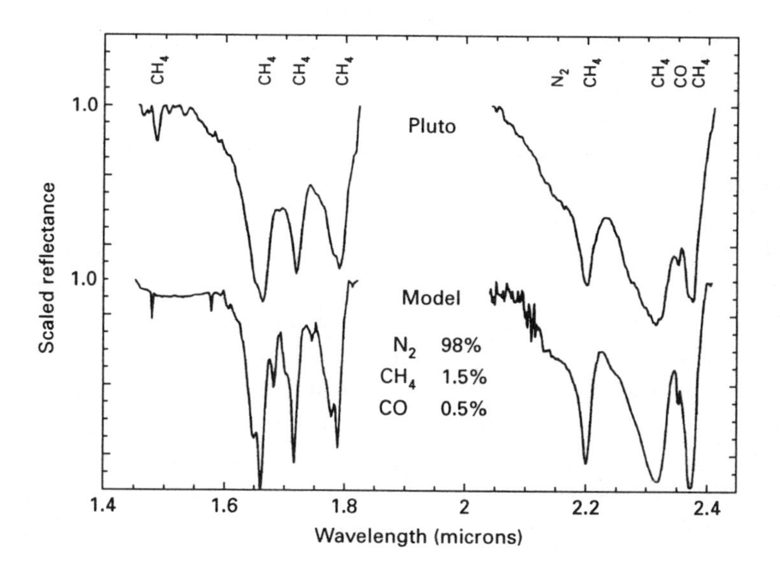

FIGURE *4.14*

Part of Pluto's infrared spectrum (top trace) compared with a theoretical spectrum calculated on the basis of a 98% nitrogen (N_2), 1.5% methane (CH_4), and 0.5% carbon monoxide (CO) ice. Note how well the theoretical model matches the observation. Some individual features in the spectrum due to the presence of the three frosts are labeled. (Adapted from T. Owen and D. Cruikshank, in *Pluto and Charon,* University of Arizona Press, 1997)

detected above the methane ice. Young's spectrum showed that the methane gas was present only in trace levels in Pluto's atmosphere. This confirmed that the methane signatures detected by Cruikshank and so many others since the mid-1970s had been due, almost entirely, to frozen methane ice, and it provided strong corroboration to the case for a nitrogen-dominated atmosphere.

Boom, it was done! Our whole view of Pluto's atmospheric composition had been turned over. The "air" on Pluto, it turned out, wasn't made of methane. Instead, it's now known to consist primarily of something so familiar—nitrogen—the major constituent of the air you're breathing as you read this page. Of course, Pluto's nitrogen is mixed with enough carbon monoxide and methane to literally choke a horse, but then there probably aren't any horses on Pluto.

Things were falling into place in rapid succession. But what about the IRAS measurements that gave temperatures of 60 Kelvin, far too high to be in agreement with the other data? When the IRAS data were re-examined, they still looked fine. When Sykes's data reductions were checked, they were bang on. When Sykes's temperature calculations were repeated, they checked. Neither Sykes nor IRAS was in error. It was our view of Pluto that, though not in error, was certainly deficient.

As it now appears, there are both relatively warm areas (those 60-Kelvin regions to which the IRAS measurements were sensitive) and relatively cool areas (the 40-Kelvin regions to which the radio measurements had more sensitivity). Apparently, the 40-Kelvin regions are the bright spots and polar caps in the mutual-event maps, and the 60-Kelvin regions are thought to be the darker regions first resolved in those maps. If this view is correct, then the bright regions should be colder in large part because they are where the nitrogen and other volatiles lie, furiously sublimating, and therefore providing an energy sink that cools these parts of Pluto's surface. Likewise, it's thought that the darker regions are warmer in large measure because they are deficient or devoid of much sublimating ice. The fact that the IRAS-derived temperature of 60 Kelvin agrees well with the radiative-equilibrium temperature calculation, described earlier for Plutonian ground devoid of ice, lends strong support to this hypothesis.

One key test of these ideas will come when IRAS's successor in space, ISO (the Infrared Space Observatory), and a new, more sensitive generation of radio-telescope instruments observe Pluto in 1997 or 1998. The observation that these instruments will perform is the measurement of how Pluto's surface temperature corresponds to the various dark spots and bright regions on the surface presented to us as Pluto rotates. If this hypothesis is confirmed, it will have important ramifications for ideas about meteorology on Pluto, implying ferocious winds, complex meteorological dynamics, and just possibly, some real weather blowing across the Plutonian landscape.

CHAPTER

5

BUILDING A BINARY PLANET

The universe is not only stranger than we imagine. It is likely stranger than we can imagine.

J. S. B. Haldane, 1926

What an intriguing little world Pluto turned out to be. Out there, far against the deep, it glides on the ragged edge of our planetary system, replete with a nearly like-sized satellite, a come-and-go atmosphere, complex surface markings, and an eccentric orbit that ranges back and forth by more distance than the entire span from here to Uranus! How did such a strange and apparently unique world come to be? How does it fit into the architecture of the solar system? The answers lie at the crossroads of the hottest areas of research in planetary science.

A GRAND DESIGN

To unravel Pluto's origin we have to begin by exploring the broader context of our solar system's architecture and how it came about.

At one level, that architecture was taught to all of us as schoolchildren. It is centered on the Sun, which it is immediately surrounded by four inner, rocky planets on tight-knit orbits. Moving farther outward, there is the asteroid belt, then the four giant, low-density gas worlds, and then, finally, lonesome little Pluto.

However, there is more to the house of Sol than meets the schoolchild's eye. There are also comets, and moons, and planetary rings. There are little pockets of debris from the era of planetary formation that orbit in lockstep with giant Jupiter. There are meteorites that have been blasted off the rocky asteroids and inner planets. And there is the gargantuan Oort Cloud of comets orbiting the Sun a thousand times farther away than Pluto itself.

Isn't it amazing to think that this inanimate collection of objects we call our solar system has any design at all? After all, why aren't the giant planets interspersed among the rocky ones? Why are the rocky planets all close to the Sun? Why is there no large planet in the gap between Mars and Jupiter? Why are the solid inner worlds so much smaller than the giants lying farther out? Why are most planetary moons 10,000 times or more smaller than their parent planet, and yet Luna so much larger in proportion to the Earth? Why are the giant planets topped with massive atmospheres? And where did all the material—the almost 500 Earth masses that are now contained in the planets—come from, anyway?

At the outset of the twentieth century, astronomers could pose such questions, but the answers remained hidden behind a veil of ignorance. Today, by contrast, we are lucky to enjoy the fruits offered by decades of intensive study using telescopes, computer simulations, probes to every planet save Pluto, and, importantly, half-a-dozen human explorations of our ancient neighbor, Luna. As a result, it is now possible to recount with some confidence how the solar system came to be as it is.

The path toward obtaining this basic, but still incomplete, understanding of our solar system's origin and evolution has been a tortured and expensive one. More than a thousand humans from places as diverse as the United States, Russia, Europe, Japan, Canada, and South America have invested their careers in it. It has cost billions of dollars, and several lives, and it is not yet done. And there is little wonder why: Because the story is more complicated, more subtle, and more convoluted than many expected it would be. Indeed, another century hence, our grandchildren, and their children, too, will still likely be unraveling the story. But that thread across generations is part of the lure of the quest. Let us see what has been learned. . . .

BEFORE

Comedian Steve Martin once said that he knew the secret to getting a million dollars and never paying taxes: "It's easy. First, you get a million dollars. . . ." Well, the secret to forming the solar system isn't so different: First, you get a universe. Now, if you want to learn about the big bang, cosmic strings, cosmic inflation, dark matter, or any of those other, well, cosmological things, then the best thing to do is to trot down to your local bookstore because you won't find them here. It's not that we don't care: We love our universe, and we love our galaxy. Some of our best friends are even cosmologists. Nonetheless, while 99% of the known universe may be dark matter or charged plasma, for our story, it's the other 1% that matters.

Our story, in fact, begins just after halftime in the history of the universe, when the Milky Way was about 4.6 billion years younger, but still mature enough to appear much as it does today. About 26,000 light years from the galactic center, out in the great spiral arms where the Sun now quietly orbits, a massive interstellar cloud of dust and gas was collapsing under its own weight, as such clouds often do. As the collapse proceeded, the cloud fragmented into clumps, many of which would become individual stars. The collapsing cloud was no doubt composed primarily of hydrogen and helium; in fact, just 2–3% of the cloud mass consisted of anything else.

As the cloud fragment with the Sun's name on it—the solar nebula—collapsed, it heated up. This heating up is due to the compression of the gas that the collapse was all about. Just as importantly, the collapse also enormously amplified the cloud's spin. Like a skater pulling in extended arms during a pirouette, the cloud spun around faster and faster as it shrank. Quickly, which in this context means in a lot less than a million years, the solar nebula formed an ever-more-intensely smoldering protostar at its center and a flattened disk of dust and gas orbiting around it (see Figure 5.1). The disk probably extended out hundreds of astronomical units from the infant Sun.

At the same time, in the little knot of galactic space within a few tens of light-years around the young Sun and its disk, hundreds of similar young stars and their disks were also forming (see Figure 5.2). Most of those stars were less massive than the Sun, but some were more massive, and some were more or less identical to the Sun. Some stars formed alone, as did the Sun, but others formed in pairs (binaries), triples, or—very rarely—even large groupings. The fraction of these stars that formed planetary systems is anyone's guess, but astronomical observations have shown that many stars get at least to the protoplanetary disk stage.

Returning to the Sun's spinning disk of material, several interesting things were taking place. For one thing, drag within the disk was caus-

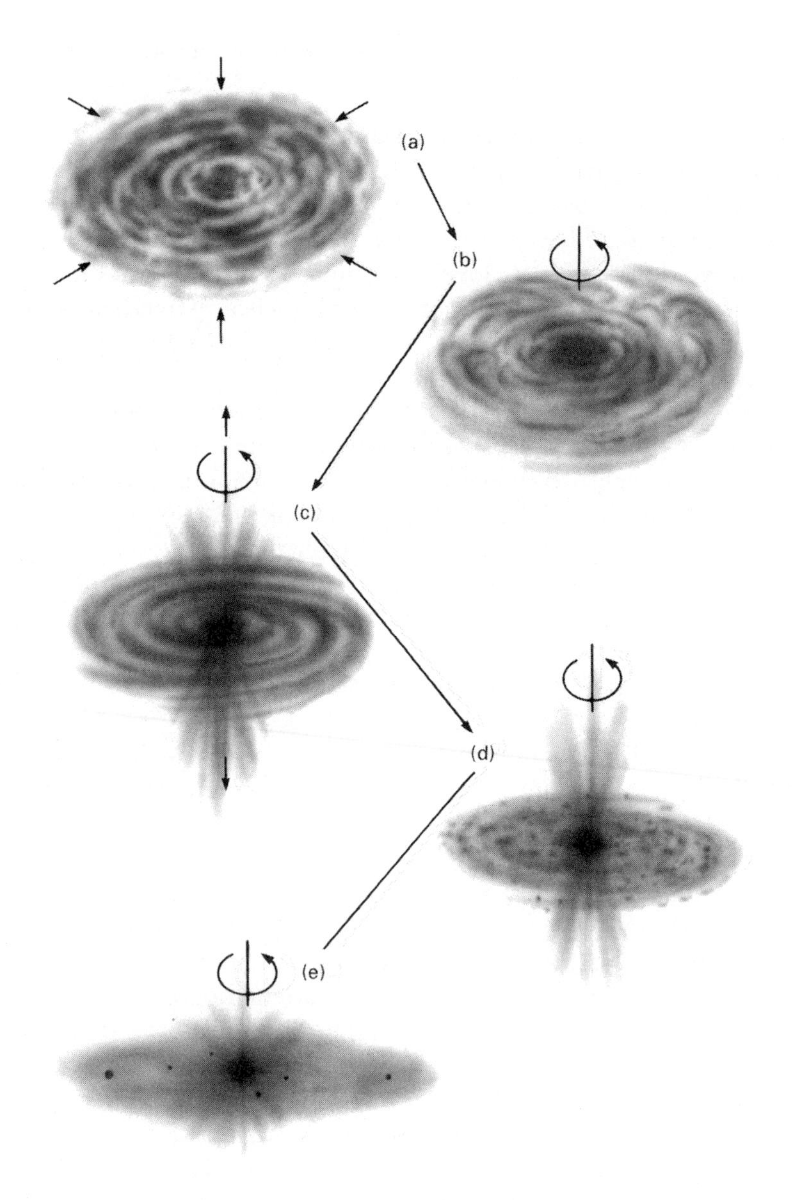

FIGURE *5.1*

The main stages in the formation of the Sun and the solar system are schematically illustrated here: (a) The cloud of interstellar gas from which the Sun and the planetary system formed begins to collapse. (b) As the cloud collapses, its rotation speeds up and it takes on a flatter shape. The center becomes hotter and denser. (c) The proto-Sun forms, surrounded by a disk of gas and dust. The young Sun blasts jets of material along the direction of its rotation axis, while more gas and dust falls into the disk. (d) Material in the disk clumps together to form planetesimals. Accretion of larger bodies takes place, along with destructive high-speed collisions. (e) The energetic wind of the young star sweeps away any remaining gas, as the major planets we know today emerge as the survivors of the formation process.

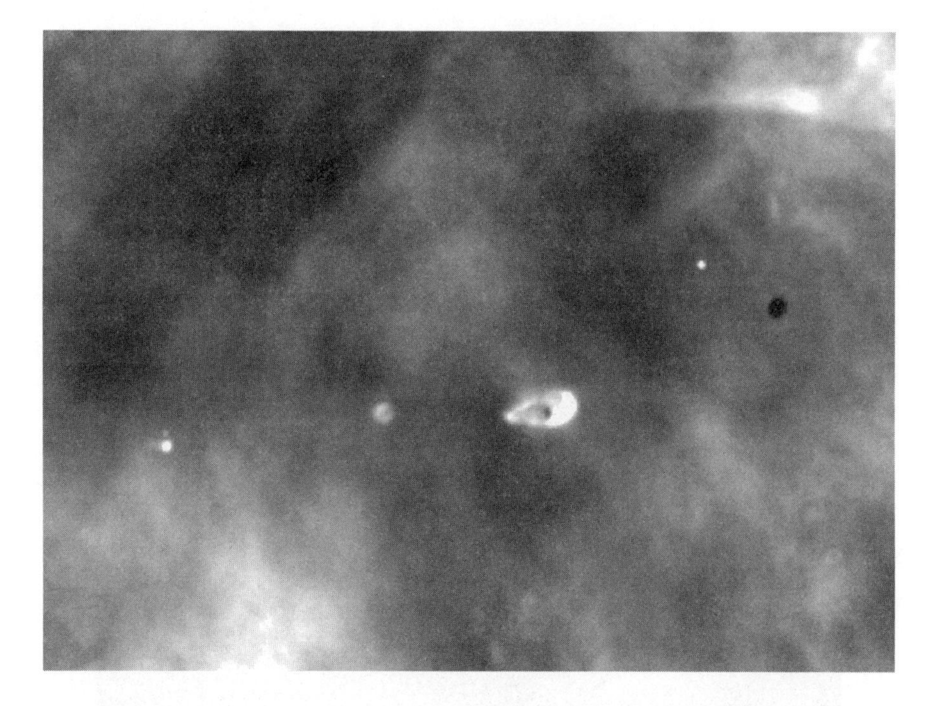

FIGURE *5.2*

A Hubble Space Telescope image of young stars in the Orion Nebula, each surrounded by a disk of gas and dust that may in time evolve into a planetary system. (C. R. O'Dell, Rice University, and NASA)

ing dust and gas to spiral inward. Some of it fell onto the Sun, adding to the Sun's mass and affecting its early spin rate. On the other hand, some of the material was funneled away in two opposite directions along its spin axis in great exhaust plumes that astronomers call "bipolar outflows." These plumes reduced the disk's mass and also helped to cool it (see Figure 5.3). Importantly for our story the dust grains within the disk were growing. Exactly how the grains grew isn't fully understood yet, but observations of star-forming regions in space support the case that somehow through low-velocity collisions or larger-scale gravitational collapse, these dirt grains, ice grains, snowballs, boulders, and ultimately, larger objects called "planetesimals" built up.

The process which began with a cloud collapsing to a disk and ended with the formation of planetesimals is a great concentrator of heavy elements. (By *heavy elements,* astronomers mean everything other than

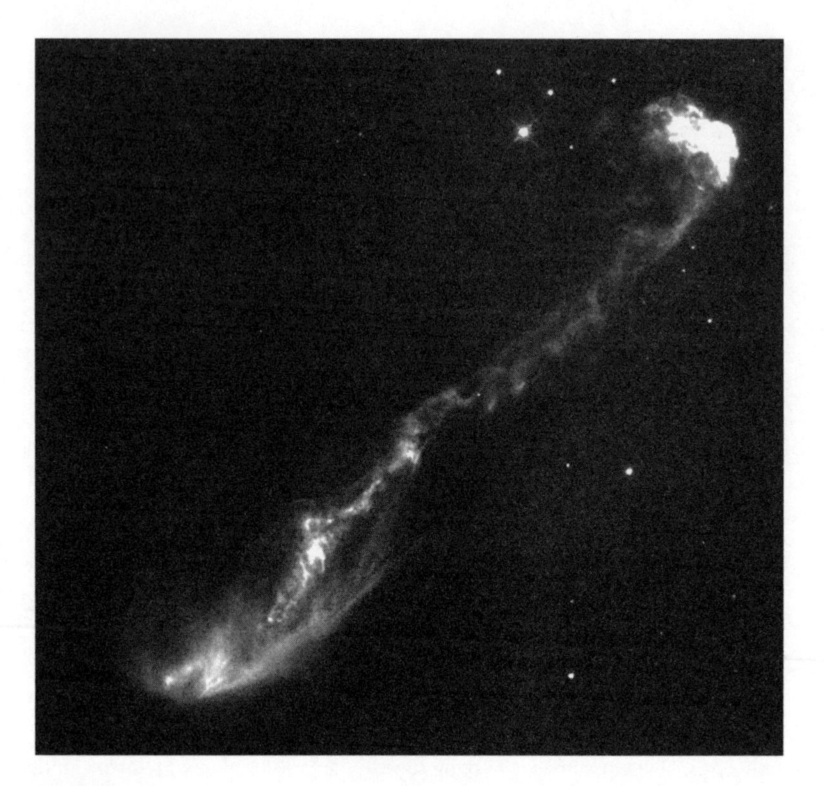

FIGURE *5.3*

A jet of gas 1½ light-years long, bursting out of the dark cloud of dust and gas at the lower left. This cloud conceals a newly forming star. The jet is known as HH47. The image was taken with the Hubble Space Telescope. (J. Morse, STScI, and NASA)

hydrogen and helium.) As a result of this concentrator effect, the planetesimals consisted primarily of heavy elements such as carbon, nitrogen, oxygen, silicon, and sulfur. The planetesimals, we know from studies of asteroids, comets, and meteorites, contained only trace amounts of helium and perhaps 10% hydrogen (primarily bound up in water ice or hydrocarbon molecules). The reason that the planetesimals were so enriched in heavier elements over hydrogen and helium is straightforward: Neither hydrogen nor helium can condense easily or are likely to become trapped in great numbers within the solid particles in the disk. As a result, most of the hydrogen and helium remained in gas form, while much of everything else, which condensed into solid ices and minerals, ended up within the planetesimals.

BUILDING PLANETS: A STANDARD MODEL

As their name suggests, the planetesimals were the solid building blocks of the planets. Close to the Sun, where temperatures were too high for water ice to survive, the planetesimals consisted primarily of rock. Farther out, probably somewhere near the place where Jupiter is today, conditions were cold enough to allow water ice to remain frozen. As a result, the planetesimals that formed at and beyond this distance were probably much like the comets we see today—fragile bergs of ice and dust. As time passed, the planetesimals grew.

To develop a picture of how this growth took place, imagine the view from the surface of a planetesimal. All around would be the thick cloud of the nebular disk. Close to the Sun, there might actually be enough light to see, but farther out, the long column of gas and dust was so thick that it cast a darker pall than any bad weather day does beneath the blackest clouds of Earth.

One of the most important processes acting in the disk was the aerodynamical interaction between solid objects and the nebular gas itself. The larger the object, the less it is affected by the gas. As a result, boulders and planetesimals weren't as strongly influenced by aerodynamic effects. However, smaller objects—such as dust grains, snowflakes, pebbles, feel greater drag,[16] and thus spiraled sunward faster than larger objects do. Aboard a large planetesimal, one would (flashlight in hand) observe a continual accretion of pebbles and rocks, along with less frequent collisions from boulders and house- and even hill-sized entrants spiraling sunward and occasionally running smack amidships. Even more rarely, another planetesimal of similar size would come looming out of the murk to collide.

Collisions are good—at least if they are gentle. For this is the way that objects grow. And fortunately, the same gas drag that causes objects to spiral inward and cross paths also keeps their characteristic collision speeds fairly low—between a few millimeters and a few meters per second.

Of course, all objects don't grow at the same rate. Bigger objects grow more rapidly because they are larger and therefore simply run into material at a faster rate than the little guys do. And planetesimals grow more quickly in the inner solar system than they do farther out. Why? Largely because at locations farther and farther away from the center of the nebula, the density of material drops, making collisions

---

[16]Because of their larger area-to-mass ratio.

progressively more infrequent. As a result, the time it took planetesimals to double in mass in the region near Earth's orbit may have been ten times less than in the region where the giant planets are.

Once a planetesimal reaches a size of about a hundred kilometers in diameter, something else occurs to help promote its growth: It becomes massive enough that its gravity begins to focus objects coming near it onto collision courses. The cognoscenti call this an "enhancement of collisional cross-section due to gravitational focusing."

Jargon aside, what matters is that once a planetesimal becomes large enough to generate appreciable gravitational focusing, its size can start to run away from others in the pack. Consequently, it grows at faster and faster rates as the mass it accretes generates more and more focusing, which in turn further accelerates accretion. If you are starting to get the feeling that the early solar system looked a lot like a kind of 1920s Keystone Cops movie, with collisions and near misses occurring at an ever increasing rate, then you're getting the right picture.

Once planetesimal seeds have grown to the point where they are about 1% of the mass of the Earth (i.e., have a diameter somewhere near 500 to 1,000 kilometers), they become large enough to significantly perturb each other's orbits as they pass one another. The cumulative effect of these perturbations can cause individual planetesimals to migrate by millions or tens of millions of kilometers (or miles, if you like). This is a small distance compared to the size of the solar system, but an enormous distance when compared to the zone within which the growing planetesimals can otherwise accrete material. It is this migration helps keep the runaways from becoming isolated in narrow little feeding zones that would otherwise eventually empty as each planetesimal eats up everything within its limited grasp.

In the inner solar system, orbits are so tightly bound to the Sun that the process of gravitational scattering can move things around, but few objects are actually ejected from the region. Virtually all of the planetesimals in the inner region where Mercury, Venus, Earth, and Mars formed had to end up either in the planets, or in the Sun.

In the outer solar system, however, the combination of less tightly bound orbits and the larger size and gravity of the planets that formed there, allowed a lot of the material to be ejected from the region altogether.

Of equal importance, the growing planets in the outer solar system could do more than just focus other planetesimals in. They were also able to capture large amounts of gas from the disk. The technical name for this process is a "hydrodynamic instability." The best mental picture for it is probably a kind of whirlpool in which gas from the surrounding nebula is funneled onto the growing planetary cores of the outer solar system at astounding rates.

Why did this process work in the outer solar system, but not the inner solar system? A key factor is that the colder conditions in the outer solar system reduced the critical mass a planet had to reach to initiate hydrodynamic accretion. Thus, whereas a planet at the Earth's distance from the Sun would have had to reach a mass of perhaps 30 or more Earth masses to initiate the rapid, hydrodynamic accretion of gas, a planet at 5 AU,[17] where Jupiter is now, only had to reach 10 to 30 Earth masses. Of equal importance, the outer solar system contained enough material for planets with masses this large to form. The inner solar system, by contrast, didn't contain enough material to make planets large enough to initiate hydrodynamic accretion.

Once hydrodynamic gas accretion begins, it proceeds very rapidly. In just tens, or perhaps hundreds of thousands of years—no time at all astronomically speaking—the accretion of gas piles on many dozens of times the mass of Earth. As a result, Jupiter's core, with a mass near 20 Earth masses, captured 300 Earth masses of gas in less time than it took the Earth itself to form! Likewise, Saturn's core captured 70 or 80 Earth masses, and the cores of Uranus and Neptune each managed to gather up a few Earth masses of gas. These four worlds would have kept on growing, too, if something hadn't intervened.

WINDS OF CHANGE

The environment in which the planets were growing began to radically transform when the Sun shed its birth cocoon and disk during an intense phase of early evolution called the "T Tauri stage." During this phase, a fierce stellar wind blew, accompanied by copious quantities of ionizing ultraviolet and x-radiation. The T Tauri stage got its name from the star in which it was discovered (T Tauri), and observations since the 1970s have shown that just about every solar-type star undergoes the T Tauri transformation by the time it is ten million years old.

This T Tauri great wind probably blew at 10 or more kilometers per

[17]Astronomers use the mean distance from the Earth to the Sun as a convenient distance scale. This distance, 149,597,870 km (about 92,960,000 miles), is called an "astronomical unit," or "AU."

second (that's 30,000 feet per second!). With this hurricane of hurricanes, virtually all of the hydrogen and helium gas in the Sun's disk was blown back into interstellar space. As a result, the remaining gas and dust around the young Sun was almost completely blown back into interstellar space. In stripping the Sun of its disk, the T Tauri wind brought light where there had been only darkness, and it left behind the swarming gazillions of planetesimals that had grown large enough to resist the wind.

You can already see in this story something that it took decades for researchers to appreciate: Timing is a critical part of the game. Why? Because it played a crucial role in determining the final architecture of the solar system.

For example, one of the critical races that took place (unbeknownst to the inanimate participants) was between the internal clock in the Sun, racing forward to the T Tauri stage, and the growth of the giant planets. If the Sun had evolved faster, or the giant planets a little slower, then the disk of gas from which the giant planets grew so fat might have been blown away before hydrodynamic accretion began. Had that happened, Jupiter and Saturn would probably remain stillborn giants, perhaps only 5 to 15 Earth masses in scale. As it happened, Jupiter and Saturn were able to accrete tremendous amounts of gas before the T Tauri wind stripped the nebula's gas away. Interestingly, it appears that farther out in the solar system, Uranus and Neptune apparently got to the hydrodynamic accretion phase so late that they were only just able to begin accreting gas when the Sun went T Tauri.

Another critical race that took place was the competition between the growth of Jupiter and the growth of planetesimals into a planet 300 million kilometers (200 million miles) sunward, just beyond Mars. Apparently, Jupiter's core grew quickly enough that the hydrodynamic accretion of gas caused the planet to rapidly balloon in mass before a significant planet could form in the region we now call the "asteroid belt." As a result, Jupiter exerted a strong gravitational influence on the region where the asteroid belt is today. As a result, collisions, which otherwise would have been gentle enough to promote growth but instead resulted in shattering. In effect, Jupiter's rapid mass growth had the consequence of terminating the growth-nurturing environment in the region where the asteroid belt is today. Consequently, the region is devoid of a substantial planet and contains only the thin debris left over from the era of shattering collisions that Jupiter induced.

What all this means is that the detailed character of the solar system is in part due to random chance—such as "which bodies" won which races. As we begin to discover and plumb the structure of other planetary systems, we will probably find that the same races turned out differently in different stellar locales, introducing a great variety in planetary systems around our galaxy.

VIOLENCE

In the 1970s, at about the same time that the consequences of the T Tauri stage of stellar evolution were becoming understood, researchers typified by George Wetherill of the Carnegie Institution in Washington, D.C. discovered something with equally fundamental consequences for planet formation. Wetherill found that making planets from planetesimals wasn't a problem. In fact, the computer simulations he constructed made lots of little planets the size of Earth's Moon and Mars. In some of Wetherill's computer simulations, hundreds of lunar-sized, and dozens of Mars-sized objects formed in the region between Mercury's orbit and the main asteroid belt. These objects, Wetherill found, were not the final set of planets that were to form, but a crowd of intermediate, transitory planetary "embryos."

Why were these embryos transitory? Simply put, when you crowd a couple of hundred small planets into the inner solar system and let them run about, they will collide and combine, which reduces their numbers. Ultimately, collisions result in a situation where only a small number of far larger planets remain, their orbits separated by enough space to isolate themselves from further collisions. One such survivor was Earth.

Wetherill's computer simulations beautifully illustrate that within a few tens of millions of years (i.e., about 1% or 2% of the present age of the solar system), the inner solar system cleaned itself up through this cascade of combinations. Different simulation runs in Wetherill's computers produce different variations on the end result, but most terminate with an inner solar system containing three to six planets with typical masses ranging from about a tenth to perhaps a few times Earth's mass.

Computer simulations by planetary-formation experts such as Al Cameron and Willy Benz at Harvard, as well as by Scott Tremaine and his coworkers in Canada and at NASA's Ames Research Center, convincingly showed that cataclysmic collisions between planet-scale bodies were common in the final stage of planetary formation. Further, these simulations showed that many impacts were so massive, and so violent, that they could melt a planet's whole surface, and at the same time, reorient its spin axis. It's thought that just such a collision stripped away much of the original mantle of Mercury, leaving behind the dense core and overlying crust that we see today. This great violence probably also caused numerous collisions that vaporized oceans or even melted the surface and deep interior of the Earth many times. In doing so, life may have lost its footing many times, as Earth's molten surface was repeatedly resterilized, but that's another story.

LONESOME DOVE

The tale of solar system formation just told was woven from almost three decades of increasingly sophisticated computer models and is buttressed by the physical evidence obtained from meteorites, lunar samples, telescopic observations of stars across the galaxy, and spacecraft visits to every major type of body in the solar system. The result is a tidy, if still sketch-work consensus about the basic path that led from the solar nebula to the rocky inner planets, an asteroid belt, and then a set of gas giants of decreasing size with distance from the Sun. The standard model also does a nice job of fitting in the comets—icy planetesimals, really—most of which were catapulted from the planetary region by gravitational action on the part of the giant planets, and which now reside in distant Oort Cloud orbits, or as flotsam among the stars. If there is something that sticks out like a sore thumb in this scenario, it's the question of where Pluto came from.

Many people have wondered, even as schoolchildren, why Pluto doesn't seem to fit into the pattern of the solar system. Why is it in that crazy orbit? Why is it out there, 10,000 times smaller than its giant planet neighbors? Why does it have a satellite half its own size when no other planet does? . . . After all, planetary formation is such a messy process. Why would the solar system create just one, little, lonely pipsqueak planet just beyond the gas giants? Astronomers wondered, too.

As it turned out, the question of Pluto's place and origin in the solar system was unexpectedly deep and profound: In coming to understand Pluto's context and origin, a dramatic paradigm shift about the content of the distant outer solar system of revolutionary proportions had taken place.

THE FAINT SMELL OF FISH

Almost as soon as Pluto was discovered, astronomers began to realize that this new world differed from the others. The three key attributes that set Pluto apart from its planetary kin were known as early as the 1930s: its cockeyed eccentricity and inclination, the fact that Pluto's orbit crossed Neptune's, and the emerging evidence that Pluto was much smaller than Neptune. Of course, no one then realized just how tiny Pluto truly was, but it did seem a different kind of cat.

In 1936, Raymond Lyttleton collected these facts together in an influential paper he published in the British research journal *Monthly Notices of the Royal Astronomical Society.* In this paper, Lyttleton outlined several possible scenarios for Pluto's origin. The one that caught on argued that Pluto was an escaped satellite of Neptune. How did it supposedly escape? Lyttleton conjectured it had suffered a close approach to another of Neptune's satellites, largish Triton, which—unlike any other major satellite of the planets—orbits its parent world on a reverse course. As shown by Figure 5.4, in the ancient gravitational exchange that Lyttleton imagined, Pluto had been ejected, and Triton's orbit had been reversed. Thus, in one tidy interaction, Lyttleton's scenario offered an explanation for (i) why Pluto is so small (it used to be a satellite), (ii) why Triton orbits Neptune backward (it had been reversed by an encounter with Pluto), (iii) why Pluto's orbit appears so rakishly disturbed (it was ejected from orbit about Neptune), and (iv) why Pluto apparently crossed Neptune's orbit. Not a bad day's work.

Lyttleton's scenario enjoyed broad public exposure and gained a wide, tacit approval from astronomy-textbook authors. But despite the fact that evidence mounted against it over the decades, as late as 1995 when Lyttleton died, most astronomy textbooks still offered his 1936 scenario as the standard explanation of Pluto's origin.

What's wrong with Lyttleton's scenario? The first problem emerged from exquisitely detailed orbital simulations. This kind of calculation is computationally intensive because accuracy demands recalculating the positions of each planet to within a few meters, tens of billions of times, and each planet's position affects all of the others. When computers became fast enough in the mid-1960s and early 1970s to accurately simulate the evolution of planetary orbits for millions of years, Pluto's strange orbit was studied closely to reveal its secrets. What researchers making these kinds of computations found is that, despite the *appearance* that Pluto's orbit crosses Neptune's, the two *never actually* come close to one another at all.

In fact, Pluto's and Neptune's orbits are locked in a repeating pattern called a "resonance," which strictly prevents close approaches. At the heart of this resonance is the amazing circumstance that Neptune's orbital period is *exactly* two thirds that of Pluto's. This repeatability

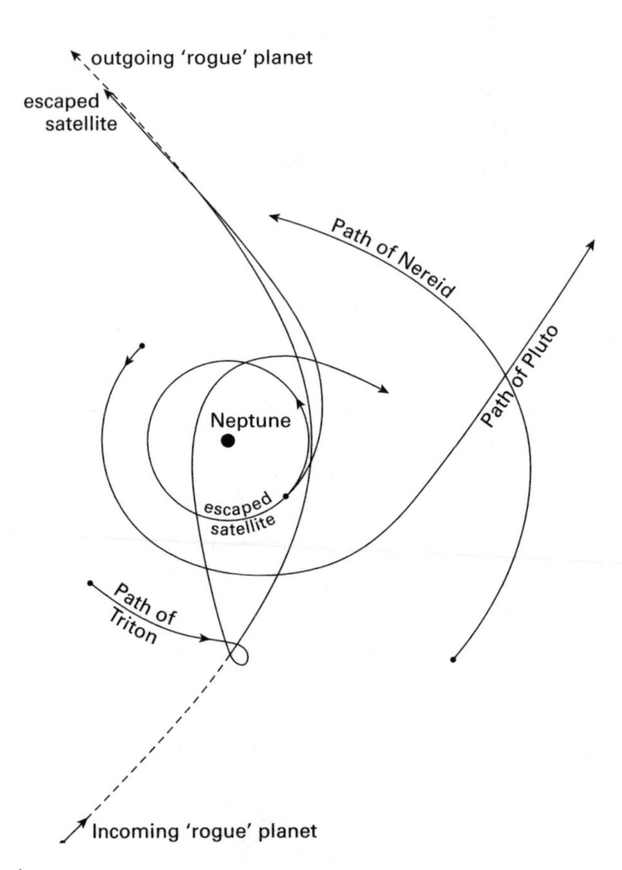

FIGURE *5.4*

After Raymond Lyttleton first proposed (in the 1930s) that Pluto may have originated as a satellite of Neptune, a number of different attempts were made to devise scenarios in which Pluto might have escaped from Neptune. The particular scenario shown here requires a "rogue planet," of about three Earth masses, to pass close to Neptune and eject Pluto. In one deft action, the rogue is shown disrupting the paths of Triton and Nereid to their unusual present-day orbits, and ejecting both Pluto and Charon completely out of their Neptunian orbits. Such a scenario is now regarded as virtually impossible, not least because the relation between the orbital periods of Neptune and Pluto ensures that the two never even remotely approach each other, and therefore that Pluto could not have originated in orbit about Neptune. (Adapted from R. S. Harrington and T. L. Van Flandern, *Icarus, 39,* 1979)

means that every time Pluto makes two laps around the Sun, Neptune makes exactly three. This synchronicity (which astronomers refer to as Pluto's "2:3" resonance with Neptune)—and the orientation of Neptune's orbit to Pluto's—ensures that the two always remain far separated. Whenever Pluto crosses inside Neptune's orbit (as, for example, it has from 1979 to 1999), Neptune lies about a quarter of the way around the Sun (from Pluto), and Pluto is at the northern extension of its inclined orbit. Although it is not known how this resonance came to be, it means that even over billions of years, Neptune and Pluto *cannot* come closer to one another than about 2.2 billion kilometers (1.5 billion miles). This is illustrated in Figures 5.5 and 5.6. What wonderful clockwork; no Swiss watch could be any finer! This alone, this celestial quarantine of Neptune from Pluto, deals a fatal blow to Lyttleton's scenario, because Pluto could not have originated in orbit about a planet it cannot approach.

As our knowledge of Pluto, Neptune, and Triton grew, Lyttleton's scenario suffered a second, equally fatal wound. Enter Bill McKinnon, a Caltech-minted Ph.D. interested in planetary origins and interiors. (Figure 5.7). McKinnon is one of those rare, brilliant mavericks, who is deeply respected both for his creativity and for his thoroughness. What McKinnon did in the 1980s was to show that Pluto's mass is both too small to have reversed Triton's orbit, and too large to have been ejected by Triton. If the orbital simulations had shot Lyttleton's hypothesis dead, McKinnon's calculations sealed the coffin.

As these Achilean weaknesses in Lyttleton's hypothesis emerged, astronomers tried a few other ideas on for size. Perhaps Pluto was an escaped member of the asteroid belt. No—how would it get all the way out where it is, and locked in that strange resonance with Neptune? Ah! Maybe it's a failed giant planet core that never accreted gas because the Sun's T Tauri clock ran faster than Pluto's growth. No, probably not—it's too small to be a giant planet core. Oh—maybe Pluto *was* a giant planet that somehow lost its atmosphere to reveal the denuded seed inside. Sure, buddy—and then why didn't Uranus and Neptune suffer similar fates?

You could almost see sweat dripping from the scientific literature to find some explanation for the little misfit planet. There was a faint smell of something fishy in the air about Pluto and its curious situation, but no one could quite put their fingers on it.

With the advantage of hindsight, it's easy to see that the old ways of thinking about this problem were simply too limited and too rooted in the need to explain Pluto as an oddity.

To explain Pluto's origin, something different was required. Some-

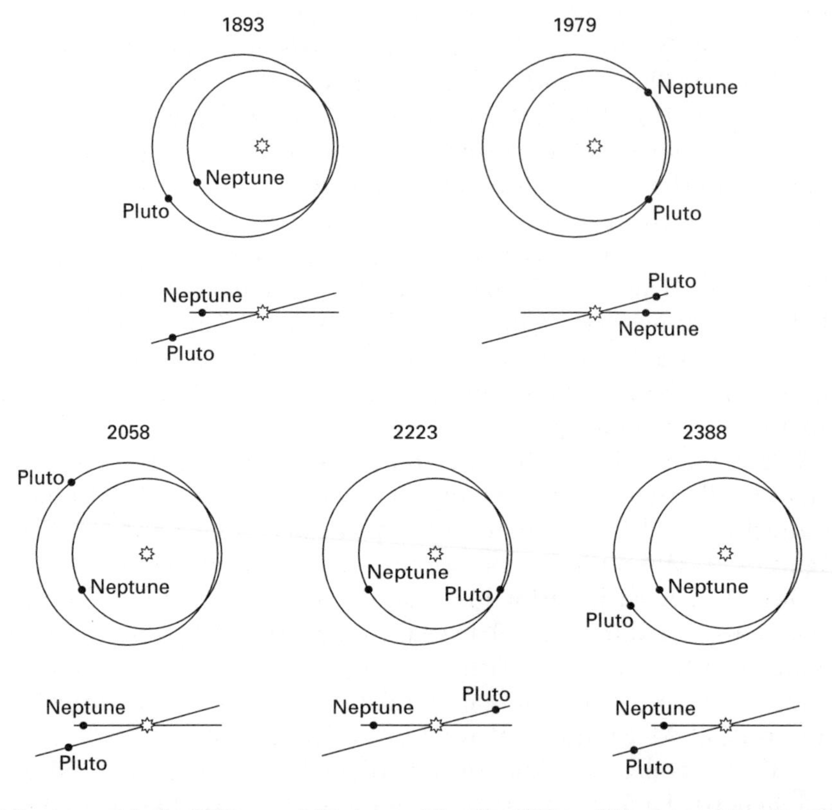

FIGURE *5.5*

Though Pluto's orbit crosses that of Neptune, the two planets are never in danger of colliding. This is because they are locked into orbits with stable periods in the ratio of 2:3 (Pluto:Neptune), which, along with the orientation of the two orbits, ensures that the two stay well apart. In fact, Pluto gets nearer to Uranus than it ever gets to Neptune. Even so, it never approaches Uranus closer than a billion miles. Upper left: Neptune and Pluto are only ever lined up on the same side of the Sun (at conjunction) when Pluto is near aphelion, as in 1893, for example. Upper right: When Pluto crosses the point where it becomes nearer the Sun than Neptune, the two planets are well separated, as seen either face on or side on to Neptune's orbit. Bottom row (left to right): Pluto and Neptune reach conjunction again only after Neptune has completed exactly 3 orbits to Pluto's 2.

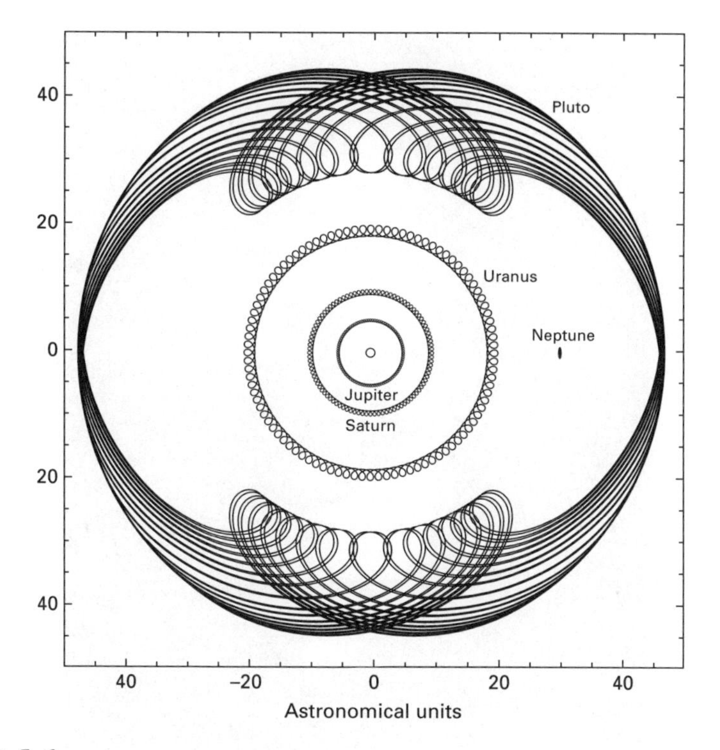

FIGURE *5.6*

The way Pluto keeps its distance from Neptune is neatly depicted in this diagram, which shows the motion of Jupiter, Saturn, Uranus, and Pluto over a 40,000-year period, from the point of view of a stationary Neptune. In this reference frame, two orbits of Pluto trace out a complete loop in just under 500 years. The figure illustrates how the relative orientation of the orbits of Pluto and Neptune slowly goes through a 20,000-year cycle (a phenomenon called "libration") but never results in a close approach of the two planets. (Adapted from R. Malhotra and J. Williams, in *Pluto and Charon,* University of Arizona Press, 1997)

thing wonderfully new. Something wholly unexpected. Something that helped to open a revolution in our whole concept of the architecture of the outer solar system. Strangely enough, the first clue came from a place far distant—the maré of Earth's Moon.

BIG BANGS

Earth's Moon had been a subject of serious scientific inquiry since Galileo first turned his telescope there at the dawning of the 17th century. And perhaps the most fundamental scientific question about the Moon investigated over the centuries between Galileo's time and ours

FIGURE *5.7*

William McKinnon: Among other things, he showed that Pluto could not have been ejected by Triton from orbit around Neptune. (J. Mitton)

has been the question of how Luna came to be. Even by the time the *Apollo* program was under way in the late 1960s, three contending theories had emerged. Each had its own proponents, but dispassionate observers of the debate agreed that none of these scenarios fit the lock with a proper key.

One of the three competing scenarios was called the "fission theory" (a kind of ultimate "windup and pitch"), in which the Moon was thought to have been shed from a hot, young, rapidly spinning Earth. But the fission scenario suffered from the fact that calculations had repeatedly shown that there was no natural way to get the early Earth spinning fast enough to fission itself into the Earth and the Moon. A second idea was that the Moon had formed as a planet in orbit around the Sun, and had somehow later been captured into Earth orbit. Unfortunately fot this idea, computer simulations of capture usually resulted in a collision between the Earth and the Moon, destroying the Moon. The third idea was that the Moon had accreted in orbit around the Earth as the Earth itself formed. But calculations couldn't explain the large mass and orbital momentum of the Moon if it formed by coaccretion. No clear favorite among the three theories emerged.

Still, in the 1960s, most scientists expected that the rich scientific harvest of samples and geophysical data that *Apollo* was expected to bring would shed enough light to make clear which of the three theories was correct, or correct enough to be patched up. But the rocks and other evidence brought back from the *Apollo* lunar expeditions didn't do this at all; instead they set the whole study of lunar origins on its ear. In fact, after *Apollo,* none of the theories looked viable! Why? In large part, this was because the proportions of different chemical elements and their isotopes in lunar samples resembled the Earth's deep mantle, rather than the material closer to the surface. How, researchers asked, could this have been the case if the Moon had spun off the Earth, as in the fission hypothesis, or if the Moon had formed either in solar orbit as its own planet or in orbit around the Earth? Just as it was with Pluto (20 years later), the old ideas about lunar formation had failed so completely that a wholly new idea was required.

Amazingly, the idea that came to triumph sprang forth almost simultaneously in two different places. Don Davis and Bill Hartmann from the Planetary Science Institute in Tucson, Arizona, hit on it first, in 1975. Al Cameron and his colleagues at Harvard thought of it only a few months later.

The Moon, these groups offered, had been dug out of the Earth in a cataclysmic collision between the early Earth and one of the Mars-sized planetary embryos that existed during the final stages of inner planet accretion. This theory, which came to be known as the "giant impact hypothesis," fit beautifully into George Wetherill's then-just-emerging view of late-stage planetary accretion. In this view, dozens, if not hundreds, of lunar-sized and larger planets were competing with one another for space and mass, and in the process of doing so, suffering mutual collisions of gargantuan scale. The idea that massive collisions contributed to the formation of the planets is widely accepted now, but it was quite ahead of its time in the mid-1970s, when most researchers still thought planets grew only by the slow sweeping up of tiny planetesimals and small debris.

Once the Hartmann and Davis and Cameron teams had described their fresh idea, rapid advances began to be made. These two teams, and dozens of others, began testing the giant-impact hypothesis against computer models, chemical evidence from *Apollo* samples, and what was known about the Earth's mantle.

Early computer simulations of the giant impact found it possible to put a lunar mass in Earth orbit, and to do so with the right orbital momentum and chemical signatures to match the real Moon—something none of the earlier lunar-origin theories could plausibly accomplish. As shown in Figure 5.8, it was also shown that such an impact could reasonably explain the difference in composition of the Earth and the Moon, and why the Moon looked so much like Earth's mantle: The cataclysmic collision had dug its building blocks out of the mantle! With these kinds of strong evidence in its favor, the giant-impact hypothesis has become *the* standard model for the formation of our Moon.

Curiously, by the mid-1980s, a very similar situation had developed with regard to explaining the origin of Pluto–Charon that had occurred with regard to the Earth–Moon system back in the 1960s. Fission didn't work; neither did capture; and neither did coaccretion. Why not? Just as with the Earth–Moon, the major stumbling block was the high angular momentum and unusual mass ratio of the Pluto–Charon pair.

It was then that something occurred to Bill McKinnon, who by then was at the University of Arizona. About the same time, the same idea occurred to Joe Burns at Cornell University, Stan Peale at the University of California at Santa Barbara, and graduate student Alan Hildebrand at the University of Arizona. Each realized that in the new ideas about lunar formation, there might also be the solution to explaining how the Pluto–Charon binary came to be. Because Pluto–Charon is an even more extreme case of a binary planet than the Earth–Moon is, a giant impact could have formed this system as well. And when pencil was put to paper to check the numbers, the details of the Pluto–Charon giant impact idea worked: Both the mass ratio of Pluto to Charon and the system's formerly inexplicable, high angular momentum fit! So convincingly did the giant impact explain what was known that fission, coaccretion, and capture quickly went by the wayside as topics of serious discussion.

Today, no clear rival to a giant impact exists for the formation of the Pluto–Charon binary. With a kind of poetic symmetry that could not have been imagined when Pluto was discovered, astronomers now appreciate that Terra–Luna and Pluto–Charon share a bond of common cataclysms that set them apart from all the other planet–satellite systems in a special, if violent way.

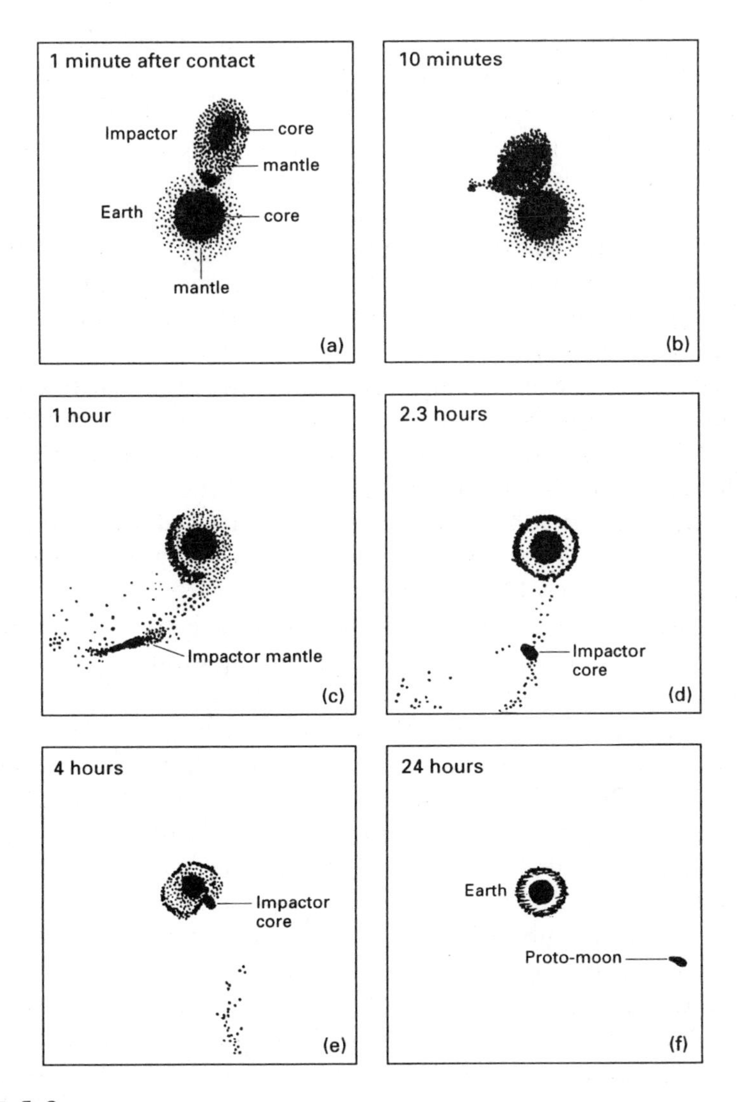

FIGURE *5.8*

A giant impact in action: The possible sequence of events if the newly formed Earth were struck by a large body (about 15% of Earth's mass) to form the Moon. Both Earth and the impactor have metal cores overlain by mantles of silicate rock. After the collision (a and b), the impactor is broken up (c), but the debris clumps together, largely through the action of gravity. The iron core of the impactor separates out and falls into the Earth (d and e). Some of the debris from the impactor's mantle accretes in orbit around the Earth to form the Moon (f). (Based on computer simulations by A. G. W. Cameron and W. Benz)

LOST FLOCK

According to the giant-impact hypothesis, the Pluto–Charon binary came to be in a scenario rather like this: First, Pluto must have formed in the ancient outer solar system. Why it was so small compared to Neptune and the other giants was not clear, but it probably had to do with Pluto's growth being cut short in the ancient past. That Pluto was a creature of the icy outer solar system was fairly certain: After all, its surface ices were too volatile to have persisted in the warm, inner regions. Even more importantly, Pluto's average density proves it contained a good deal of water ice, which, of course, the rocky inner planets and asteroids do not.

Then, sometime after Pluto formed, another object, with a mass of perhaps one-tenth to one-half that of Pluto's (and a diameter near 1,000 kilometers), struck Pluto. The impactor couldn't have been much smaller, or else it wouldn't have packed the punch to create the binary. In the aftermath of the blow, enough material was stripped from Pluto's surface and into orbit to create Charon. From that material, Charon accreted in orbit around Pluto.

This was a nice scenario, and it worked far better than Lyttleton's old idea, but it begged several important questions. For example, where did the impactor come from that collided with Pluto to trigger Charon's formation? How did the impact just happen to be the right kind to create a binary? (A gentle collision would instead have created a merged world, and a sufficiently violent one would have shattered both Pluto and the intruder to smithereens.) So, too, why would the outer solar system make just one, highly unique, tiny binary on its outer border? After all, the outer solar system looked pretty empty—where would the impact have come from, and what were the odds it would have struck little Pluto?

The way to approach he last of these questions is to estimate the amount of time it would take Pluto and a proto-Charon in a similar orbit to run into one another. Time scale questions like this can be answered in a rough, but robust way, by a method called a "particle in a box (PIB) calculation." The PIB theory was originally developed in the 19th century for the molecular theory of gases, but it can be applied in many other situations as well, and it's a common tool used by astronomers to estimate time scales.

The PIB equation for calculating the time scale for object A to be hit by object B only requires knowing three things about the situation: (1) how big the target (object A) is, (2) what the concentration of projectiles B is, and (3) how fast the projectiles are moving about, relative to the target. Given these three numbers and a hand calculator, we can quickly find the approximate time it would take for such a collision to become likely. Let's see how the factors that go into this calculation are determined.

At first brush, it might seem that the target size for our situation would just be the size of Pluto. However, it's more realistic to recognize that a collision occurs whenever the target and the projectile touch. Therefore, if the projectile is a good fraction of the target's size (as a giant impactor must be), then the target size is better represented by a sum of Pluto plus the impactor. To improve this estimate a little further, we should also account for the fact that Pluto's gravity would attract the incoming projectile. Astronomers call this gravitational focusing.

To estimate the concentration of projectiles (i.e., how many of them were packed per cubic kilometer), the simplest and most conservative thing to do is to assume that there was just one impactor in all the volume of space that Pluto's orbit draws out over time. That's about one object per 30 billion billion *billion* cubic kilometers. To calculate the speed at which the projectile and target stumble around, blindly searching for one another, we need to know their orbital eccentricities. Because it's impossible to know what the eccentricity of the giant impactor's orbit actually was, we can just solve the PIB equation for the whole range of plausible encounter speeds, so that we know the answer to the rough time after which a collision is expected, no matter what the relative speed of Pluto and the fateful impactor actually were.

In 1990, a PIB calculation just like this one was done for the first time. It's amazing that it had never been tried before. After all, if you're going to hang your hat on a giant impact for making the Pluto–Charon binary, then it seems you'd better know whether it's likely or not. In any case, it hadn't been done.

Plugging in the numbers gave an astonishing result. Regardless of whether the impactor and Pluto had a low or a high relative speed, it's hard to make them collide in a reasonable amount of time. The space is just too big for them to find one another very quickly.[18]

The specific implication of the PIB calculation was this: If Pluto and a proto-Charon were permitted to wander about in the deep outer solar system in orbits like Pluto's, they would take between 10,000 and

---

[18]Of course, we're using terms that suggest the two objects were consciously hunting for one another, when in fact, they were simply inanimate shots rattling around in an imaginary box defined by their orbits.

10,000,000 times the age of the solar system to collide![19] Put another way, in the age of the solar system, the odds of Pluto being hit by a single giant impactor orbiting in the same region of the solar system were between 1 in 10,000 and 1 in 10,000,000. That dog won't hunt.

How could this be reconciled with the convincing case that a giant impact is needed to create the Pluto–Charon binary? The obvious answer is to put more than just one impactor in the mix. With more projectiles running about, Pluto would thus get hit more often. As it turns out, to make the collision likely in less than the age of the solar system, the mathematics dictate that between 300 and 3,000 Charon-forming impactors had to be roaming about in the early days of the solar system.

When this was first suggested in 1990, the thought of it was a little daunting. On the one hand, if the early outer solar system contained only Pluto and the Charon impactor, then our best hope for explaining this binary planet (the giant-impact hypothesis) had a ridiculously low probability of having actually occurred. On the other hand, if the impact did occur, then it implied that Pluto wasn't born a lonely little footnote to planetary formation but is instead the telltale survivor of a formerly *huge* population of similar-sized objects that were present in the ancient outer solar system.

Amazingly, the choice between these two world views was about to become very clear.

---

[19]This range of uncertainty is enormous, primarily because the range of possible encounter speeds was left to be just about anything consistent with objects orbiting the Sun. But if you think like an astronomer, you realize it doesn't matter. Regardless of the encounter speed, the time needed to have a high probability of a collision between Pluto and a single proto-Charon is far, far longer than the age of the solar system.

CHAPTER

6

ICEFIELDS AND ICE DWARFS

Where have all the flowers gone, long time passing?
ECCLESIASTIES

If Pluto was the only smoking gun to suggest the presence of large numbers of now-missing objects in the ancient outer solar system, the story might have stopped right there. After all, the formation of the outer solar system was so rough and tumble that some very unlikely things might very well have happened. However, while the Pluto–Charon binary could be a fluke, it's a bit much to accept that the freak would be the only survivor of the melee. Taking a page from paleontology, when a strange new fossil is discovered, researchers have to assume that the discovery is a relic of a formerly widespread species, as opposed to a mutant that just happened to survive the ages.

Researchers wondered what other evidence supported the case for a vast population of ancient ice dwarfs. The first support came from studies of how the planets got their tilts. As it turns out, the answer appears to be through giant impacts. By the late 1980s, it had become pretty well accepted that the tilts of the inner planets came from Wetherill's crashing and combining of lunar-sized to Mars-sized bodies into the growing rock worlds. It was also appreciated that Jupiter and Saturn don't have large tilts because they are just too massive to be rocked by anything much smaller. But what about Uranus and Neptune? The spin axis of Uranus is tilted 98 degrees from an upright position, and Neptune is tilted over almost 30 degrees. It isn't easy to tilt these spinning tops: Each has a mass about 15 times that of Earth (see Figure 6.1). The only really viable way to generate these tilts would be for Uranus and Neptune to have swept up some very large competitors to their respective thrones during accretion. Large is almost an understatement here. To tilt massive Uranus and Neptune, the ancient impactors must have been nearly Earth-sized or larger.

Astronomers can use the PIB method described in Chapter 5 to estimate how many large impactors would have been needed in the

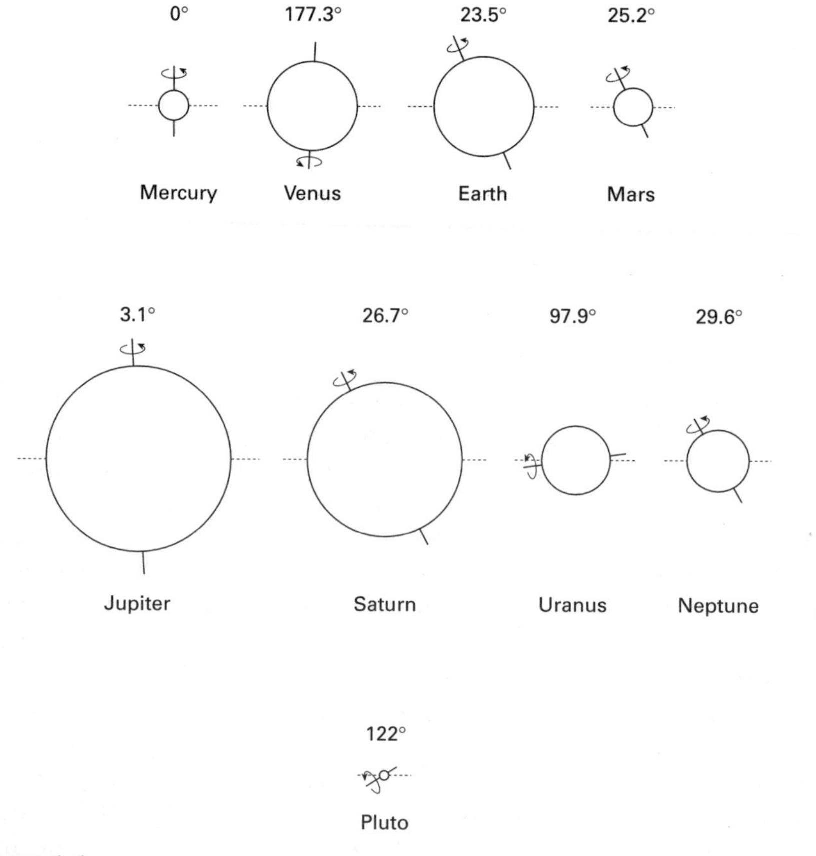

FIGURE *6.1*

The degree to which the rotation axis of each planet is tilted with respect to the plane of its orbit around the Sun. A tilt greater than 90 degrees means that the planet's rotation is *retrograde*—that is, in the opposite sense to its motion around the Sun. The planets in our solar system with retrograde rotations are Venus, Uranus, and Pluto. (The size of the giant planets are shown scaled down by a factor of 5, relative to the terrestrial planets.)

Uranus–Neptune zone to make it probable that Uranus and Neptune ended up with the tilts they have. Such a calculation reveals that there must have been between 10 and 100 roughly quarter-Earth-mass or larger objects in the region where Uranus and Neptune formed, in order to have a significant chance of tilting Uranus and Neptune to their present orientations. This is very strong evidence that the outer solar system was more crowded in its early days than it is now.

Yet another line of evidence comes from studies of Neptune's largest moon, Triton. Triton (see Figure 6.2) is similar to Pluto in several striking respects. For one thing, Triton's 2,700 kilometer (1,690 mile) diameter is just a little more than 10% bigger than Pluto's (2,300 kilometers, or about 1,450 miles). Triton and Pluto also each have densities of almost precisely twice that of water. Each also displays a surface covered in nitrogen, carbon monoxide, methane, and water ice, with

FIGURE *6.2*

A view of Neptune's moon Triton, assembled from 14 separate images obtained by *Voyager* 2: This mosaic shows a variety of different kinds of terrain. The streaked south polar cap is apparent at the bottom. The dimpled surface visible toward the upper right has been likened to the skin of a cantaloupe melon. To the right there is a region of smooth plains and low hills, which is the most densely cratered part of Triton seen by *Voyager.* (NASA)

nitrogen dominating the mix. In fact, planet Pluto and satellite Triton are more similar to one another than are any two planets in the solar system. Another amazing thing about Triton is the fact that its orbit around Neptune is backward, relative to Neptune's rotation and to the orbits of all of Neptune's other satellites. This retrograde orbit is dead-on evidence that Triton was gravitationally captured by Neptune, and therefore that Triton was once a planet in orbit about the Sun.

Triton's capture from solar orbit has been studied by some of the best planetary theorists in the world, including Peter Goldreich and Bill McKinnon. The trick in finding a way to get Neptune to capture Triton is the need to bring Triton close to Neptune while dissipating excess orbital energy and momentum, so that Triton enters a stable orbit around Neptune.

One way to effect Triton's capture is a perfectly aimed, grazing collision by Triton through Neptune's atmosphere. (Not too shallow, or Triton would escape. Not too deep, or Triton would never reemerge into orbit.) Alternatively, Triton might have just happened to have struck a satellite of Neptune while passing through the system, shattering the luckless moon, but in the process reducing its own energy and momentum enough to achieve orbit about Neptune. A third alternative envisions that the capture was aided by gas drag from the tenuous nebula around Neptune during its formation.

What's clear from these ideas is that regardless of whether it entered Neptune's atmosphere just right, just happened to hit a satellite of Neptune, or passed by during the brief period when Neptune's atmosphere was grossly extended, something quite improbable occurred. And just as was the case with the Charon-progenitor finding Pluto to make the binary, the best way seen to avoid the long odds of the improbable event is to postulate that many "Tritons" must have once been extant in this part of the solar system.

Putting together the evidence from the Pluto–Charon binary, the tilts of Uranus and Neptune, and the capture of Triton, it seems that the outer solar system is telling us something. At Uranus, at Neptune, and at Pluto, there are the smoking guns of long-past events indicating that the ancient solar system made small and moderate-sized planets in great numbers.

Were this not the case, the odds of creating the tilts of Neptune, *and* the presence of Triton in a retrograde orbit, *and* the Pluto–Charon binary would be too long to take seriously.

In analogy to the gas giant nomenclature, astronomers often refer to the suspected population of small planets that apparently once inhabited this region as the "ice dwarfs." And given all of the circumstantial arguments pointing toward their former presence, the issue isn't whether the ice dwarfs existed: It is where they are now. To answer this question, we first have to understand how the outer solar system became empty.

WHERE HAVE ALL THE PLUTOS GONE?

For more than 20 years now, computer simulations have been used to study the final stages of outer-planet formation. The most notable of these formation simulations were performed by Victor Safronov and his colleagues in the Shternberg Institute in Moscow and by the team of Julio Fernández in Montevideo, Uruguay, and Wing Ip in Lindau, Germany.

These two groups have demonstrated some important things about the formation of the giant planets. First, they have shown that because the giant planets are sufficiently far from the Sun, and so massive, that they gravitationally scatter (i.e., eject) far more planetesimals from their formation zones than they actually accrete. A key reason for this is simply that objects are more likely to pass close to (but miss) a planet than to actually hit it. Objects on such near miss trajectories receive a strong orbital deflection[20] causing them to be ejected from this region of the solar system. With literally trillions of planetesimals orbiting between Jupiter and Neptune, the melee around each budding giant planet must have been tremendous, with every manner of rock, boulder, collisional shard, and planetesimal that passed sufficiently close being ejected.

Once the giant planets began reaching their present-day masses, there was virtually no refuge in the outer solar system from this effect. In fact, studies of the orbital dynamics of the outer planet regions show that within just 10 million to 100 million years after reaching their present-day masses, the giant planets emptied the whole region, from four to almost 34 AU, of everything but one another.[21] What wasn't emptied on this time scale was cleared out in the first billion years after the giant planets formed.

Interestingly, Jupiter and Saturn are so massive that almost all of the objects that they scattered were ejected completely out of the solar system. However, because Neptune and Uranus have far smaller masses than Jupiter and Saturn, recent simulations indicate they deposited be-

[20]As did the *Voyager* and *Pioneer* spacecraft that flew by the giant planets.

[21]The fact that the giant planets themselves remain is due in part to their wide spacings and in part to their large relative masses.

tween 10% and 30% of the planetesimals and debris that they scattered into distant, weakly bound orbits beyond Neptune.

Most of the planetesimals and smaller debris launched to these distant orbits by Uranus and Neptune now populate a zone that stretches a thousand times farther away than the giant planets. This region is called the "Oort Cloud," after the late Dutch astronomer Jan Oort, who first recognized its existence in 1950. This gigantic cloud of debris from the era of giant planet formation may contain 10^{12} (a million million) ejectees a kilometer or more in size.

As the outer planetary region emptied, everything went, including the ice dwarfs. Nothing stood a chance in the celestial tug of war against the giant planets. Although the ice dwarfs had diameters of hundreds to a few thousand kilometers—and masses a million times that of comets and the small planetesimals, to the giant planets, the ice dwarfs were gnats. Like the smaller debris, most ice dwarfs were probably ejected into interstellar space. A small fraction, however, must have been deposited in distant, stable orbits in the Oort Cloud, where they should still reside today.

Of course, this all begs a question about Pluto and Triton. Given that Uranus and Neptune were so efficient at gravitationally sweeping up or sweeping out everything between and around their orbits, why do Pluto and Triton remain? The answer to this question was recognized about ten years ago—Triton and Pluto survive *only* because they are resident in stable *dynamical niches,* where they are protected from ejection. These niches are the astronomical equivalent of the LaBrea tar pits in California. Whereas the tar pits preferentially preserve old saber-toothed tiger bones that would otherwise have been long lost to erosion, the dynamically stable niche regions of the outer solar system preserved long-extinct relics from the formation era of the giant planets. Triton's niche is its tightly bound orbit around Neptune. Pluto's niche is its 2:3 resonance that prevents close approach with Neptune. Had Triton and Pluto not found safe havens, the giant planets might have erased all the evidence of the ice dwarfs that once inhabited the outer solar system.

THE EDGE

When the concept of a thousand or more ancient ice-dwarf planets first hit the streets as a scientific paper in the spring of 1991, reaction was mixed. Some thought it just the right answer to the long-standing lack of context for Pluto and Triton (see Figure 6.3). Others thought it little more than a nice but speculative theory, unsupported by observa-

FIGURE *6.3*

The cover of the weekly American science magazine *Science News,* on September 21, 1991, proclaimed the new idea that Pluto may once have been just one in a whole population of icy dwarf planets that formed in the early outer solar system.

tion. Why? Well, if the theory was right, then the very evidence to confirm it had been neatly removed, billions of years ago. That is, according to the ice-dwarf scenario itself, all of the other ice dwarfs must be gone by now. "How convenient."

Even for scientists, seeing is the key to believing, so the question of whether these objects truly existed has come down to finding concrete evidence of the multitudes of ice dwarfs that were thought to have formed along with Pluto. Looking to the far away Oort Cloud didn't seem promising. Calculations showed that a Pluto-like object ejected into the Oort Cloud would be a billion times fainter than Pluto—so faint that even the very largest telescopes could not hope to find it. Worse still, such an object would only move one arcsecond every four years, making it almost indistinguishable from motionless stars.

There had to be an easier way, and it came when it was suggested that some of these objects might be found in another stable region that might lie as close as a billion miles beyond Neptune.

The region beyond Neptune, which astronomers call the "trans-Neptunian zone," has been a frontier begging for exploration since Kenneth Edgeworth (in 1943 and 1949) and Gerard Kuiper (in 1951) wrote seminal papers pointing out the same great puzzle of the region. What intrigued Edgeworth and Kuiper was the sharp border that the solar system presents at 30 AU, where Neptune orbits. Inside 30 AU, there are the planets, but beyond 30 AU, there appeared to be, well, . . . nothing but tiny Pluto.

Why should that be? Why, Kuiper and Edgeworth asked, should planetary formation end so abruptly? Why didn't it instead peter out? Why, they asked, wasn't there at least a debris field of planetesimals left over from outer-planet formation? Why was there only Pluto?

Edgeworth and Kuiper's ruminations were on the trail of something big, but very few midcentury astronomers appreciated that. In part, this was because the technology really didn't allow their question to be easily addressed. It was also that, in 1950, the *lack* of material beyond Neptune just didn't strike very many people as very odd. Edgeworth and Kuiper felt that the abrupt, almost perfect emptiness beyond the last planet was bothersome, but they couldn't drum up much enthusiasm or interest about it among the larger astronomical community.

The beginnings of paradigm shifts are often like that. In fact, Edgeworth and Kuiper's early musings are reminiscent of the rumblings in late 19th-century physics: Sometime around 1890, it was commonly said that physics was about completely solved, and that once the photoelectric effect and the Michelson-Morley speed-of-light experiments were explained, physics, as a field, would be all but complete. Little did the purveyors of this notion know that relativity, quantum mechanics, particle physics, fusion, and a host of other discoveries lay just over their horizon. So, too, the astronomical community of the 1940s and 1950s didn't realize that the trans-Neptunian region was a whole new frontier for planetary science—full of surprises that better instruments would eventually reveal.

THE HUNTERS

As we just noted, Kuiper and Edgeworth wrote down their ideas about the puzzle they felt the empty trans-Neptunian region presented, but the technology of the times prevented much from coming of their speculations. So, like the two men themselves, their ideas soon passed into history.

So, too, time passed—over four decades of it, in fact. There came and went a cold war, and several hot ones. There came and went *Apollo,* the *Mariners,* the *Vikings,* and the *Voyagers.* There came and went Elvis, James Dean, the Beatles, and Kurt Cobain. There came commercial jet transport, weather and communications satellites, and a computer revolution that continues without apparent bound. Throughout, the question Edgeworth and Kuiper posed remained completely unresolved, and still largely unappreciated—for over 40 years.

The place, as it often is in modern astronomy, was atop Mauna Kea Mountain, nearly 3 miles above the shores of the Pacific, on the big island of Hawaii. The players (see Figure 6.4) were Dave Jewitt and his former graduate student Jane Luu, who was at the time a postdoctoral researcher at the University of California at Berkeley. Jewitt and Luu wanted to answer Kuiper and Edgeworth's question. Their tool was a 2.2-meter diameter telescope, with almost exactly the size of the Hubble Space Telescope, but located on the ground. Their search had begun in the late 1980s, with the advent of a new generation of sensitive CCD cameras that could obtain images of the sky with a clarity and depth few had imagined possible even in the early 1980s.

Jewitt, a British expatriate and professor at the University of Hawaii, was 35 years old. Luu, a former Vietnamese refugee who, as a girl literally escaped in the last wave before the fall of Saigon, was still in her twenties. When he was still a Caltech graduate student, Jewitt had used the telescopes atop Mauna Kea and a then-state-of-the-art CCD camera to spot the faint nucleus of Comet Halley returning toward the Sun before anyone else did. In the intervening decade since that 1982 discovery, Jewitt, with students like Luu, had been pushing CCD cameras further and further in the study of faint objects in the deep outer solar system. During that time, CCDs had also grown larger, making it possible to capture a broader search swath with every image, and more sensitive as well, largely as a result of engineering advances that reduced their electronic noise.

Jewitt and Luu weren't alone in undertaking the search. Accomplished astronomers such as Hal Levison and Martin Duncan, Anita Cochran, and Tony Tyson were trying the same thing. All of them knew it wouldn't be easy. Tombaugh's search for Pluto had long previously proven that nothing as bright as Pluto (15th magnitude) was out there, moving slowly, against the stars. And in the 1970s, astronomer Charlie Kowal had conducted another, deeper search, this time pene-

FIGURE 6.4

David Jewitt and Jane Luu, who discovered $1992QB_1$, the first Kuiper Belt object to be identified. (J. Mitton)

trating to almost 19th magnitude—100 times fainter than Pluto. But despite years of effort and a search that wrapped halfway around the ecliptic, Kowal found only one object, Chiron (see Figure 6.5), and it was orbiting primarily between Saturn and Uranus. At the end of his search, Kowal estimated that there might at most be a handful more objects 100 or so times fainter than Pluto, but anything else out there—if there was anything else—was fainter, and would be significantly harder to find.

With that knowledge in hand, the various trans-Neptunian search teams began their deep surveys at the end of the 1980s. All four groups used the strategy of searching the opposition point pioneered by Tombaugh 6 long decades before. Initially, each group reported negative results, meaning they had found nothing, and would have to search more sky, and to fainter limits, if a discovery was to be made. Some groups dropped out. Others persisted in the search. None worked harder than Jewitt and Luu.

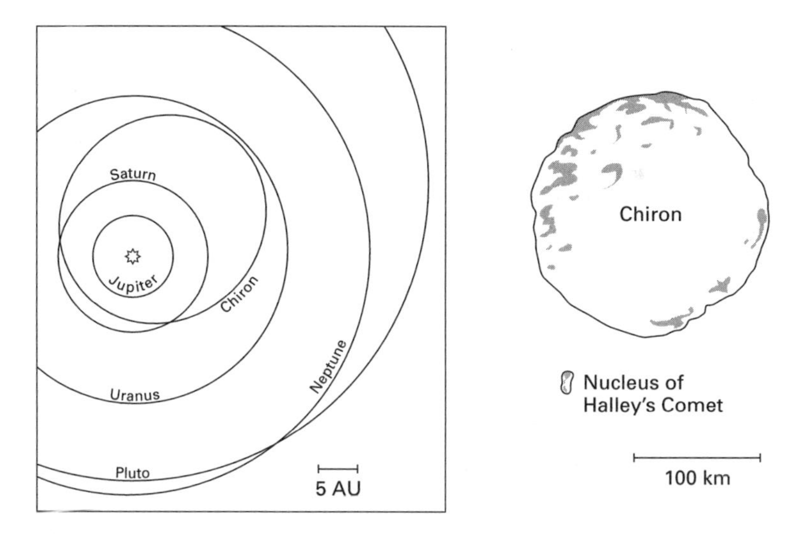

FIGURE *6.5*

Chiron, now considered the prototype of a class of solar-system bodies known as "Centaurs," seems to be intermediate between a comet and an ice-dwarf planet such as Pluto. Chiron's orbit relative to the outer planets is shown on the left. On the right, we show that Chiron is much larger than a typical cometary nucleus (the nucleus of Halley's Comet is shown for comparison). Nevertheless, Chiron develops a diffuse coma around itself—a characteristic associated with comets rather than with planets or asteroids.

SMILEY

Then, in late August of 1992, after four years of work, Jewitt and Luu were rewarded with the detection of a faint little ember, slowly moving against the myriad blazing stars on their CCD images. The object was *thousands* of times fainter than Pluto, and more than 50 times fainter than anything Charlie Kowal could have detected in his search 15 years before.

Like Tombaugh more than 60 years before, Jewitt and Luu wanted to check their work before releasing the result. Surefootedness, they knew, is far superior to the embarrassment of a highly public retraction. So, over the next few days, before the Moon came back to wash out the night sky and halt their work, the pair gathered additional observations of the tiny patch of sky surrounding their find. Their repeated imaging showed that the object was clearly real and, importantly, crawling across the ecliptic star fields at just the rate expected for a trans-Neptunian object.

One of the first people whom Jewitt and Luu informed was Brian Marsden, of the Harvard–Smithsonian Observatory in Cambridge, Massachusetts. Marsden is one of the most careful and meticulous orbit plotters in the world, and in his role as Director of the IAU's (International Astronomical Union's) Minor Planet Center, Marsden was the one to judge when the distant object's reality was sufficiently well-established to become public knowledge.

Jewitt and Luu determined the distance to their prey from its rate of motion on the sky. The faint, slowly moving object they had found was most likely about 43 astronomical units (AUs) from the Sun, and some two billion kilometers (1.2 billion miles) beyond Neptune, but the uncertainty on this distance estimate was large. Jewitt and Luu found that if they assumed that the object's reflectivity was low, as expected for its ancient, radiation-battered surface, then the object's diameter must be in the neighborhood of 260 kilometers (140 miles). That's more than 10,000 times the size of Halley—and like Kowal's Charon. At this size, the nomad was probably big enough to become a kind of lumpy sphere, indicating it was near the stage at which objects reached the clear hallmark of planethood: roundness. No doubt about it, the distant wanderer could be called a minor—or better yet—a dwarf, planet.

Marsden evaluated the data Jewitt and Luu had collected, and on September 14, 1992, he issued an e-mail bulletin and telegram called an "IAU Circular" to astronomers around the world. Such bulletins are issued several times a month to rapidly inform the worldwide astronomical observing community of discoveries that would benefit from rapid follow-up observations (e.g., supernovae or new comets). Marsden was careful to couch his remarks about the trans-Neptunian discovery in guarded language that would alert other astronomers of the find without overstating the case that a trans-Neptunian object had actually been confirmed. Marsden's announcement read, in part:

> *D. Jewitt, University of Hawaii, and J. Luu, University of California at Berkeley, report the discovery of a very faint object with very slow (3 arc seconds/hour) retrograde near-opposition motion, detected in CCD images obtained with the University of Hawaii's 2.2-m telescope at Mauna Kea. The object appears stellar in 0.8 arc second seeing, with an apparent magnitude* R = 22.8±0.2 . . . *Computations . . . indicate that 1992QB$_1$ is currently between 37 and 59 AU from the Earth but that the orbit (except for the nodal longitude) is completely indeterminate. Some solutions are compatible with membership in the supposed Kuiper Belt, but the object could also be a comet in a near-parabolic orbit.*

As Marsden's announcement indicated, there was one more hurdle to pass before Jewitt and Luu could be sure that their find (Figure 6.6) was indeed a true inhabitant of the trans-Neptunian region, first postulated 40 years before Gerard Kuiper. They had to eliminate the possibility that the object could be on an elliptical orbit plunging toward the Sun through the trans-Neptunian region, rather than permanently circling there. This would require repeated monitoring of 1992QB$_1$'s movements over a period of three to six months.

According to the strict nomenclature rules of the IAU, Jewitt and Luu's object, called "1992QB$_1$," could not be named until its orbit could be precisely established. 1992QB$_1$ is not a very romantic name for an object of such potential import,[22] but IAU nomenclature rules for the discovery of objects in the solar system clearly state that a more personal pedigree would not be accepted until the object's orbit was well established. The reason for delaying assigning a name at this stage is to prevent named objects from being lost, owing to poor knowledge of their orbits. Jewitt and Luu knew they would have to wait for formal naming until the orbit was determined well enough to satisfy the IAU, but they told anyone who asked that they had already selected the

---

[22]By the way, the IAU's nomenclature code is this: year of discovery, followed by a letter for the 2-week period in the calendar when it was discovered, followed by a letter (and if necessary then a number) to indicate the order of discovery in that 2-week period. Thus, 1992QB$_1$ was the 28th new object found in period 17 (i.e., Q) of 1992. (The 27 others were more mundane discoveries in the asteroid belt.)

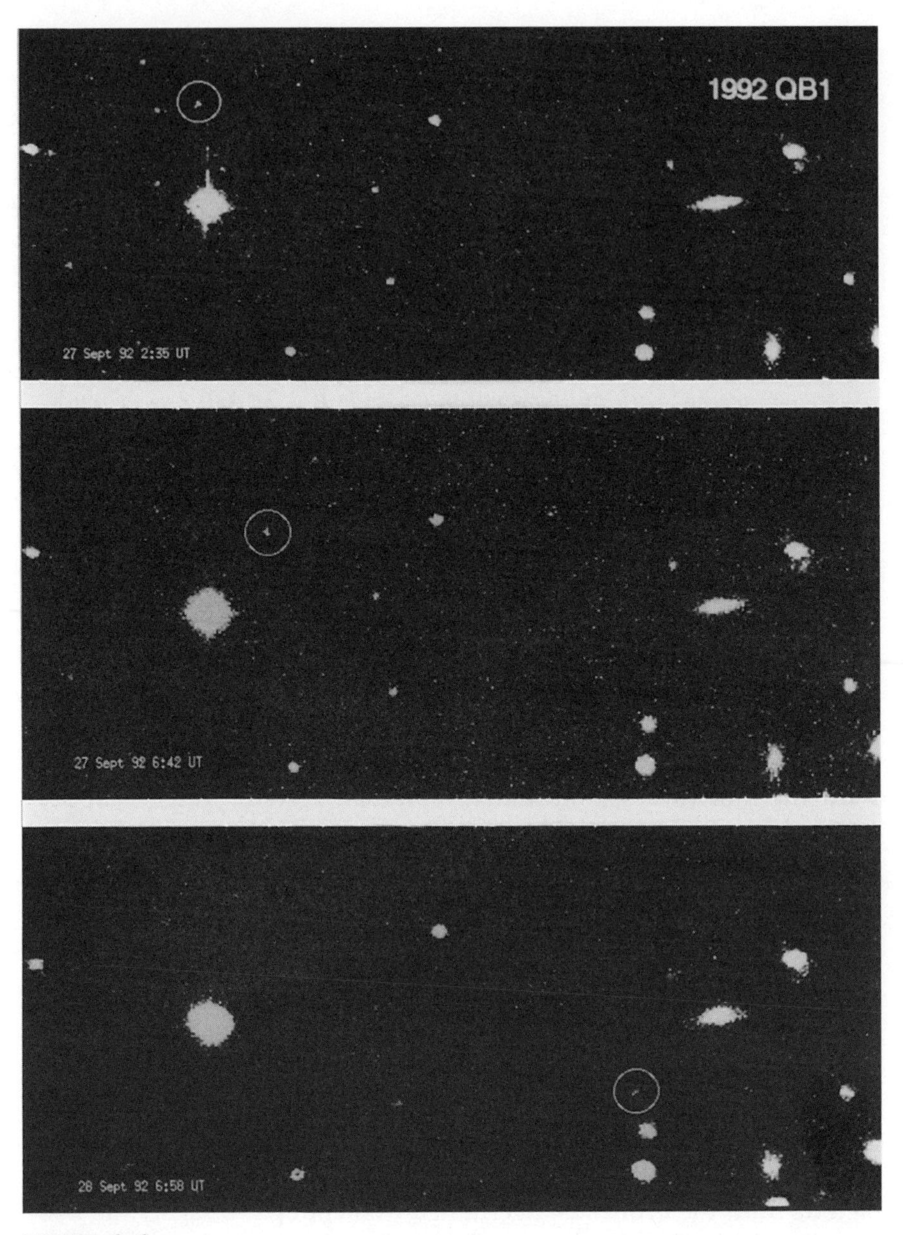

FIGURE *6.6*

This series of three images of the Kuiper Belt object $1992QB_1$ was made on September 27 and 28, 1992, about one month after its discovery. The images were taken with the European Southern Observatory's 3.5-meter New Technology Telescope in Chile. The faint (twenty-third magnitude) image of $1992QB_1$ is circled on each frame. It was moving by about 75 arcseconds per day against the stars. (European Southern Observatory)

name they planned to give their find. It would be *Smiley,* after their favorite spy from George Le Carré's popular novels.

PLUTO'S KIN

Despite the fact that its orbit had not been confirmed to be in the Kuiper Belt, the news of 1992QB_1 moved rapidly from astronomers to the astronomical press to the larger news media. Within a day of Marsden's IAU Circular, "QB_1" was on the front pages of the *New York Times, The Times* of London, and dozens of other newspapers around the world. Before the month was out, the story was on CNN and in *Time* magazine. The firestorm of coverage was so intense that for a while, some astronomers actually expected Jewitt and Luu to find their way into *People* magazine.

Finding QB_1 opened up a floodgate of other discoveries in the trans-Neptunian region. As Jewitt and Luu worked to get additional observations and improve knowledge about QB_1's orbit, they continued their search for other trans-Neptunians. Less than 6 months after QB_1 was announced, Jewitt and Luu found a second trans-Neptunian, designated 1993FW. See Figure 6.7. Then, in the late summer of 1993, they found two more: 1993RO and 1993RP. And the same week, a team from the United Kingdom, led by Iwan Williams, announced they had also begun detecting trans-Neptunians, with the discoveries of 1993SB and 1993SC. By late 1994, the count had reached 15, and by late 1977 well over 50 of Smiley's long-lost brethren had been nabbed. A few looked to be as big as 600 kilometers in diameter, large enough to swallow 1992QB_1 several times over.

What was the most stunning about the rapid pace of discoveries of Kuiper Belt objects was that the dozens of QB_1-like discoveries being made were just the tip of an iceberg. Consider why. All of the hundreds of images taken in the search for these QB_1s covered only a few square degrees of sky, which is just a minuscule fraction of the thousands of square degrees near the ecliptic where such objects are to be found. Comparing the area that had been searched to the area of the entire ecliptic region, it became clear that some 70,000 Smiley-like objects must be orbiting in the region between 30 and 50 AU from the Sun.

Further, when a team of four astronomers, led by Anita Cochran of McDonald Observatory, turned the Hubble Space Telescope in 1995 to look for objects smaller and fainter than could be seen through the Earth's shimmering atmosphere, they discovered evidence for hundreds of millions of Halley-sized comets in the trans-Neptunian re-

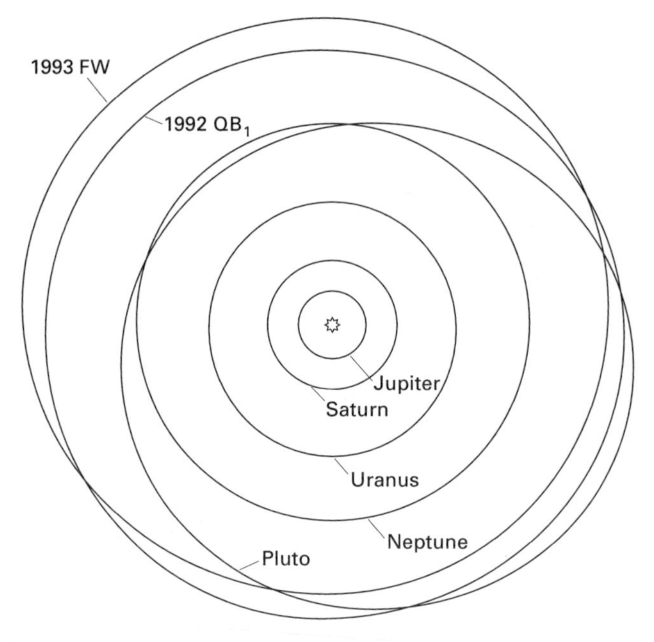

FIGURE *6.7*

The orbits of the first two trans-Neptunian objects (other than Pluto) to be identified: $1992QB_1$ and 1993FW.

gion.[23] Not only was the trans-Neptunian region littered with dwarf planets, but it was also teeming with comets and smaller debris. Pluto was not alone after all! In fact, as shown in Figures 6.8 and 6.9, it seems to be orbiting in among a veritable zoo of smaller kin left over from the era of outer-planet formation.

One particularly important fact about the origin of the Kuiper Belt, and of Pluto, is that there doesn't seem to be enough mass in the region to build objects anywhere near the size and mass of Smiley (not to mention Pluto, a thousand times larger still) in the age of the solar system. What this finding clearly implies is that the QB_1s and comets seen there today are just the remnants of a formerly much larger population of objects that was present in the distant past, before Neptune grew to its present size, indicating that a dramatic clearing of the ancient trans-Neptunian region took place.

In fact, the most recent work on the clearing of the trans-Neptunian

[23]This was just as theorists such as Martin Duncan, Scott Quinn, and Scott Tremaine, and later others had predicted from their studies of cometary orbits in the late 1980s.

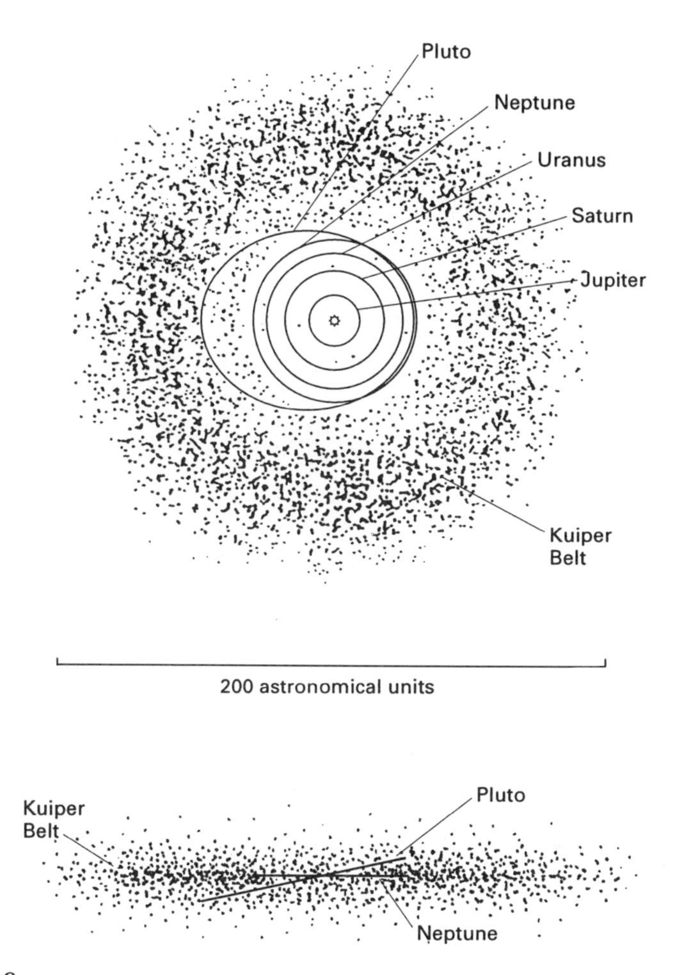

FIGURE *6.8*

The Kuiper Belt (or Disk) is thought to occupy a region extending approximately from the orbit of Neptune to a distance of 100 astronomical units (AUs) from the Sun. It is shown here schematically face on and from the side.

region shows important parallels to the clearing of the asteroid belt by Jupiter. In both cases, the clearing involved both orbital perturbations and a dramatic increase in collision velocities, which converted a formerly growth-nurturing environment to a Darwinian shooting gallery pervaded by high-speed collisions that stopped growth dead in its tracks. As a result, both the asteroid belt and Kuiper Belt appear to be regions where a larger planet might have grown, had it only won a race against the giant neighbor Neptune we see today.

FIGURE *6.9*

Gerard P. Kuiper (1905–1973), an early Pluto pioneer after whom the Kuiper Belt is named, was one of the most distinguished planetary scientists of his time. (Yerkes Observatory)

The best estimates of the combined mass of all the objects in the trans-Neptunian zone today (including comets, QB_1s, intermediate-sized bodies, and Pluto), is about one quarter to one half of an Earth mass. By contrast, simulation results suggest that some 10 to perhaps 50 Earth masses of material were originally needed in the region in order for objects hundreds of kilometers in size, like the QB_1s, to have accreted in less than the age of the solar system. If that was in fact the way the early outer solar system was, then the abrupt edge that Kuiper and Edgeworth first noticed at Neptune wasn't originally an edge at all, but simply something that evolved over time as Neptune depleted the trans-Neptunian region, and the QB_1s and Pluto represent the remnant clues to the ancient era when the whole trans-Neptunian region was far less empty than it is today.

FULL CIRCLE

With the advantage of hindsight, we can now see that Pluto's discovery in 1930 presaged the discovery of the vast debris field of the Kuiper Belt that occurred in the 1990s. Because Pluto was (and is) so much larger, and therefore so much brighter than any of its present-day kin, it *appeared* to be alone in the yawning vacuum beyond Neptune.

For 60 years, Pluto's small size, unusual orbit, and apparent isolation in the region beyond Neptune, fooled us into trying to explain Pluto as an oddity, when in fact it was a beacon, showing the way toward a then-unseen flock of smaller objects quietly plying the backwoods of the planetary system beyond Neptune.

In the 1970s, Pluto's clockwork resonance with Neptune and its double-planet nature became known, thereby precipitating a crisis in understanding its origin. These facts eliminated all of the Lyttleton-like Pluto-as-a-freak theories, in favor of much less ad hoc and far more exciting possibilities. That shift in thinking began in earnest when Charon's size was realized, in turn forcing new thinking about how a double planet could be made, and so sparked the idea of a giant impact for Pluto. The giant impact, in turn, led to the idea that the ancient outer solar system was littered with lots of Plutos and mini-Plutos, the ice dwarfs.

Then, when the technology of CCDs enabled deep enough searches to discover Pluto's family, Pluto's true relationship to the outer solar system was finally seen clearly. The discovery of the Kuiper Belt washed away forever the old problem of explaining that lonesome little Pluto. In point of fact, we now see that Pluto was probably built in a region containing the building blocks of at least one more large planet, which never formed, and that the key to Pluto's small size is probably the influence of nearby Neptune, the growth of which induced a gravitational stirring of the nest where Pluto and its cohort embryos were growing. That stirring resulted in the collisional battering that whittled away the massive reservoir of material that once populated the 30–50 AU zone, and it rendered stillborn the dwarf planets that were growing toward full planethood in this region of the solar system. Pluto and the smaller remnants of that population that we observe today serve as testimony to the vagaries of planet formation, and they provide a laboratory for us to prove what planet formation looks like when arrested in midstride.

Although we are sure now that Pluto formed amid a vast population of smaller bodies, the main thing we don't know today is how common Pluto- and Triton-sized objects were in those early days. It seems that something like at least a hundred, and perhaps a few thousand objects this size and larger were needed if Uranus and Neptune were to be tilted over, Triton captured, and the Pluto–Charon binary created.

Still, none of the trans-Neptunian objects found to date are much larger than 20% of Pluto's diameter. Perhaps we will find the missing cohort. Perhaps we won't. Perhaps another good idea is needed. The situation is unfolding before our very eyes.

CHAPTER 7

EVEREST

They say you want a revolution, well, you know, we all want to change the world.
THE BEATLES

With this chapter, we turn a final corner, moving away from a description of what things have been learned about the Pluto–Charon binary, to the planning for a spacecraft mission to reconnoiter this fascinating pair of frozen frontier worlds. Before we move on to the subject of a possible mission to Pluto, however, it is useful to remember what past spacecraft missions to planets have taught us, and to summarize what we know—and what we yearn to know—about Pluto and Charon.

CRADLE VISTA

Perhaps the main lesson of planetary science is just how deeply limited the field was before space exploration became a reality. The limitations of that era were not the size of the telescopes, but a series of other factors. First, there was the comparatively primitive nature of the instruments and detectors available to put at the focus of ground-based telescopes. (By some accounts, the advances in instrument and detector technology make each telescope a hundred thousand times more effective today than in the 1950s and 1960s.) Second, the computers to analyze the data and model the results were scarce and (when available) equally primitive. Third, astronomers were blind to some of the most valuable parts of the electromagnetic spectrum, including the rich terrain of the ultraviolet and infrared regions. Fourth, there was the stifling turbulence of the Earth's atmosphere, which reduced resolution so far below the theoretical limit of the telescopes themselves that the planets appeared only as fuzzy orbs and their satellites as simple pinpoints. Without the ability to even *see* what the surfaces of the planets looked like at moderate resolution, planetary astronomers were less well off than even the proverbial blind man describing an elephant.

Given the severe limitations of mid-20th century astronomical tools and techniques, is it much of a mystery why the solar system we know today seems so very different from the one that Tombaugh, Kuiper, and colleagues knew at mid-century? Yes, *their* solar system consisted of nine planets, but more than half were so poorly understood that even the lengths of their days were uncertain. Further, they knew not a single definitive fact about the composition of any planetary satellite, save about our Moon and Saturn's Titan. So, too, *no one* knew whether any, all, or some of the planets (except Earth) had magnetic fields. And the rings of Jupiter, Uranus, and Neptune were *so* unsuspected that reputable astronomers criticized cartoons for showing rings around planets as a common occurrence. Similarly, circa the dawn of the Space Age, some responsible astronomers thought Mercury had a substantial atmosphere, and that Venus was quite plausibly habitable. Others imagined Mars might be teeming with an alien biology, and that the asteroids might be an exploded planet. Pluto was widely thought to be a rogue misfit . . . . The list goes on, but we shall not.

The solar system known to the generation of astronomers and enthusiasts who toiled in the years between the Second World War and the dawning of the Space Age was as far removed from the solar system we know today as a 1957 Chevy is from a 1997 Corvette.

The scientific revolution brought about by the exploration of the planets has most fundamentally been a revolution in perspective. That exploration taught us that the solar system is far more diverse, more subtle, more complex, and more beautiful than anyone suspected. Exploration taught us that the richness of nature isn't limited to rain forests or summer skies, and it taught us that most of our cherished, armchair understandings of the planets and their retinue of lesser worlds were but a child's view of the larger world.

AN ORNAMENT AGAINST THE DEEP

Although we know that remote exploration by telescope often underestimates the richness and intricacies of Nature, it is useful to summarize what has been learned so far about Pluto–Charon. Much of what we know has come about because we now have the technological tools to probe an object so faint and far away, but much as well has come through the good luck of having recently had both the rare mutual events and a once-every-248-year perihelion passage to observe. It is from those good fortunes that a stick-figure portrait of the Pluto system which we now summarize has been taking shape.

To begin, Pluto is not just another planet with a satellite but is instead a binary planet, consisting of two comparably sized but otherwise vastly different worlds: Pluto and Charon. This strange binarity, more than any single attribute (save possibly its distance from the Sun), characterizes the Pluto–Charon system.

The Plutonian pair orbit the Sun together every 248 years, coming as close as 29.5 AU, and receding as far as 49.6 AU. Their orbit around the Sun is locked in a clockwork resonance with Neptune, which indicates that Neptune played some important role in Pluto's past, though it is not yet clear precisely what that role was.

The sizes of Pluto and Charon are now known to an accuracy of a few tens of kilometers. We also know that the pair lies on its side, spinning on their axes every 6.3872 Earth days. Charon orbits Pluto in Pluto's equatorial plane, with this same period; as a result, Charon hovers over one spot on Pluto's equator. Charon never sees the other side of Pluto, and so, too, that far side is ignorant of Charon's face.

All searches for other satellites in the Pluto–Charon system have failed to find anything. These searches have placed strong limits on the mass and size of any satellites that might still remain undiscovered. Nothing larger than 10% of Charon's diameter—a mere 100 kilometers or so—could be hiding there, still unknown to us. There is the very real chance that Pluto has no other companion save Charon.

The Pluto–Charon binary probably formed through a giant collision billions of years ago. Based on the odds of such a collision occurring, and the discoveries of so many objects in the Kuiper Belt region where Pluto orbits, it's now widely accepted that Pluto and Charon are among the largest remnants of a once-teeming population of miniature planets called "ice dwarfs" that were formed during the era of giant-planet building in the deep outer solar system.

Let us now look a little deeper at the two bodies that make up the Pluto–Charon binary. We start with the smaller of the pair, Charon.

It is known that Charon doesn't show very much variation in brightness as it rotates, which indicates that its surface appearance is blander than Pluto's. Charon's gray color and much darker albedo than

SOME BASIC ATTRIBUTES OF PLUTO AND CHARON

Parameter	*Pluto*	*Charon*
Rotation period	6.3872 days	6.3872 days
Approximate diameter	2,360 km	1,200 km
Density	$\approx$ 2.0 g cm^{-3}	1.8–2.2 g cm^{-3}
Surface reflectivity	55%	36%
Lightcurve amplitude	38%	8%
Known surface ices	CH_4, N_2, CO, H_2O	H_2O
Atmosphere	Confirmed	Doubtful

Pluto's also contribute to Charon's blander appearance. Spectroscopic studies have revealed that Charon's surface is covered with water ice; other ices, or even rocky materials, may lie on Charon's surface as well, but they haven't been detected yet. Similarly, there isn't any proof of (or even much indirect evidence for) an atmosphere on Charon. Based on everything that is known about Charon, which isn't much more than described here and in the accompanying table, the smaller member of the Pluto–Charon binary is expected to be rather like the similarly sized, gray, water-ice-covered satellites of Uranus.

Now we turn to Pluto, which is the life of the Pluto–Charon party. Twenty years of spectroscopic studies of Pluto's surface show that it is covered with a complex mélange of exotic snows, including methane (CH_4), carbon monoxide (CO), and nitrogen (N_2). Although the nitrogen ice dominates Pluto's surface composition today, the proportions of the various ices change both with Pluto's seasons and with its back-and-forth swings in distance from the Sun.

Pluto has an atmosphere, which envelopes it in a puffy shroud thousands of kilometers high. At the surface, the pressure of the atmosphere is just a few, or perhaps a few tens, of microbars—less than 1/10,000 the pressure atop Earth's highest peaks. Pluto's atmosphere, like Earth's, consists mostly of N_2 gas, but unlike Earth's atmosphere, Pluto's also contains significant quantities of CO and CH_4. The CO and CH_4, though present only in trace amounts, play an important role in generating the temperature structure of the atmosphere as it rises above the surface. There is evidence that the atmosphere may have haze layers, and that it waxes and wanes in mass as Pluto moves around the Sun.

Pluto's atmosphere must also contain some as-yet-undiscovered trace constituents that result from the chemistry of sunlight shining on the N_2-CO-CH_4 mixture. A similar kind of chemistry acting on the surface ices darkens and reddens Pluto's surface. As a result, Pluto's highly reflective surface sports a tinge of ruddy pink.

Pluto's density is known well enough to constrain its internal composition. It is almost certainly a mixture with 60–70% rocky materials, a few percent exotic ices like those detected on the surface, and perhaps 30–35% water ice. We suspect that Pluto has differentiated into a layered planet, but that cannot as yet be proven.

Pluto's brightness varies dramatically as the planet rotates, which provides strong evidence of a highly variegated surface. (It's no wonder, with such a complex mixture of surface ices and their contaminants!) These suspicions were first confirmed by the maps made using data from the mutual-event season that lasted from 1985 to 1990.

Owing to the intricate patchwork of bright and dark ices and other materials on Pluto, its ground temperature appears to vary from place to place by huge proportions—perhaps by as much as 50% from the warmest (~60K) to the coldest (~35K) spots. Such large temperature gradients, in turn, probably drive intense winds and complex circulation patterns. No doubt these too shift with the seasons, as the ices migrate and the atmospheric bulk readjusts to the rise and fall of solar heating caused by Pluto's motion around the Sun.

And there, unfortunately, our knowledge comes to a rather abrupt end.

BEHIND THE CURTAIN

The picture of Pluto–Charon that we were just able to sketch has about as much detail as astronomers might have painted of Mars in 1965 or Io in late 1977, just before each was visited by reconnaissance spacecraft. In a single day the first spacecraft flybys of Mars and Io revolutionized our knowledge and perception of those worlds, just as did the first close-up scrutiny that spacecraft have provided of Mercury, Venus, Europa, Triton, Neptune, and more than a dozen other worlds.

The most indelible lesson of planetary science is that Nature always outperforms our expectations and, in doing so, often shatters Earth-based paradigms on pain of close-up inspection. With Pluto, the lesson remains the same: We must remove the curtain that distance imposes, in order to finally embrace and understand the reality of the world. *Nothing substitutes for being there.*

Still, Pluto's distance from the Sun makes reaching it about as difficult a technical challenge as the solar system presents. Just consider what its great distance implies: long flight times to a dim and far away place, where solar cells won't work, and communications are a thousand times more difficult than at Mars or Venus. No wonder it's been said that Pluto is the Everest of solar-system exploration.

Everest or not, the astronomers who spent the late 1970s and then the 1980s discovering what a fascinating and important planet Pluto is weren't daunted. As the 1980s began to slip away into history, a small but determined band of Pluto's best and most dedicated explorers decided to grasp the bull by the horns and try to bring about a space mission to reconnoiter Pluto and Charon. They knew that it would not be any easy task—for the great ship NASA turns slowly, at best, even in recent times.

GATHERING

A Chinese proverb states that even the longest journey must begin with a single step. The long journey to Pluto began at a red-checkered table in a small Italian restaurant in downtown Baltimore, one cool evening in mid-May of 1989. A dozen planetary astronomers, fresh from a symposium summarizing the advances made in the study of Pluto during the 1980s, were gathered together. The focus of their dinner conversation was: how to motivate NASA to undertake a reconnaissance mission to Pluto. The founding members of the "Pluto Underground," as this group sometimes called itself, included Fran Bagenal, Rick Binzel, Marc Buie, Bob Marcialis, Bill McKinnon, Ralph McNutt, Bob Millis, Alan Stern, Ed Tedesco, Larry Trafton, Larry Wasserman, and Roger Yelle.

The central problem, the 12 agreed, was that the *Voyager* (and earlier *Pioneer*) reconnaissance missions to the deep outer solar system had left Pluto off their travel itinerary, out in the cold, so to speak. This fate had been sealed in 1979 when the *Voyager* project elected to skip Pluto in favor of a more in-depth exploration by *Voyager 1* of the Saturn system, and a daring plan to send *Voyager 2* on to Uranus and Neptune instead of Pluto (which was off in another direction). The decision that *Voyager 1* would make a close flyby of Titan and concentrate on getting the most bang out of its Saturn encounter would preclude it from visiting any other planets. Why? Because the Titan flyby would put the spacecraft on a post-Saturn trajectory that drove it away from all three of the more distant planets.

In defense of that regrettable decision, it is true that, back in 1979, little was known about Pluto, so its scientific value wasn't appreciated.

By contrast, the alternative, a close approach to Saturn's planet-sized moon, Titan, and the Uranus–Neptune trip offered a sure bonanza of returns, and a higher probability of success (since the spacecraft wouldn't have to survive the extra five years to reach Pluto). Right or wrong, the die was cast. Neither *Voyager* would visit Pluto.

By 1989, however, so much more was known about Pluto that many planetary scientists were beginning to regret the *Voyager* decisions made a decade before. Worse, by 1989, NASA's planetary program was stuck in a programmatic cul de sac, funding only giant, second-generation missions such as the $2 billion *Galileo* and *Cassini* outer-planet orbiters, and the $1 billion *Mars Observer.* Reconnaissance was out. Expensive, deep-exploration missions were in.

How would the young Pluto Underground contingent manage to redirect the inertia of NASA's planetary program to reconnoiter Pluto? They weren't sure themselves, but they believed strongly that the exploration of the last planet would have the public appeal to propel itself forward . . . if NASA would only recommend it to Congress and the Administration for approval. To accomplish that, they knew that the planetary-science community in the United States would first have to reach a consensus that, in a political environment where new NASA missions were rare, a Pluto mission deserved approval.

Over pasta that spring evening, the members of the Underground agreed to begin by organizing a letter-writing campaign to Geoff Briggs, the director of NASA's Division of Solar System Exploration at NASA Headquarters. Their first goal was to raise the flag of Pluto exploration so high that it could not be ignored.

CLIMBING

The same summer that the Pluto Underground was organizing itself, something wonderful fell into their lap: a shift in scientific opinion that led to dramatically wider scientific support for a mission to Pluto. Ironically, it was *Voyager* 2 that ignited this change. By 1989, *Voyager* 2 was a battle-hardened veteran of encounters that had revealed the secrets of more than a dozen worlds. Now, *Voyager* 2 was making its exit from the planetary system via a daring encounter with Neptune and its bizarre system of satellites.

As it turned out, however, Neptune wasn't the star of *Voyager*'s final flyby. It was Neptune's largest satellite, Triton, that stole the show. *Voyager*'s close pass to Triton (see Figure 7.1) revealed it to be one of the most interesting worlds of the solar system. The scientific excitement

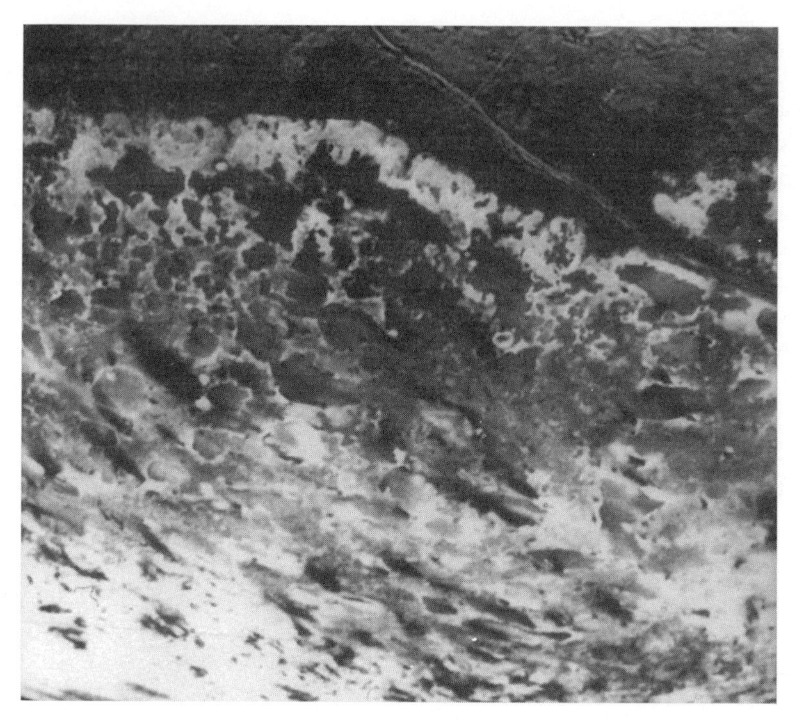

FIGURE *7.1*

Among the planets and moons of the solar system already imaged at close range, Triton is probably the object most similar to Pluto. This *Voyager* close-up of terrain near Triton's south polar region shows numerous dark plumes or streaks, which originate at very dark spots. These are thought to be vents where gas laden with dark particles has erupted into the nitrogen atmosphere from below the surface. The plumes have then been drawn out by winds. The dark streaks have been deposited on the polar terrain of bright frost, mottled with darker areas. (NASA/JPL)

generated by Triton, which was known to be the best analogy to Pluto in the solar system, motivated NASA's Briggs to action.

BASE CAMP

Less than 60 days after *Voyager 2*'s Triton flyby, Geoff Briggs organized and founded a Pluto-mission study, called *Pluto 350*. This study got its name for the arcane reason that the spacecraft would have a mass target of 350 kilograms (770 pounds) . . . barely half the mass of *Voyager*. The goal of the *Pluto 350* work was to determine whether a workable scientific mission could be put together in a package this small. Given the trend of 1970s and 1980s planetary-exploration missions, which featured multiton spacecraft, a good number of experienced insiders simply found the idea of such a small spacecraft flying across the entire solar system to be, well—ridiculous. The members of the Pluto Underground didn't much care. Perhaps it was their youthful enthusiasm, or their youthful naivete, but as far as they were concerned, planetary spacecraft were sinking under their own weight, and a small spacecraft mission to the smallest planet was just what the doctors ordered.

Briggs put Robert Farquhar at the helm of the *Pluto 350* study and funded a design team at NASA's main planetary exploration center, the Jet Propulsion Laboratory in Pasadena, California, to work for Farquhar. Briggs also asked Pluto Underground members Stern and Bagenal to serve as the scientific liaison to the mission, through one of NASA's science working groups.

Farquhar (Figure 7.2), who was then a senior civil servant at NASA's Goddard Space Flight Center in Greenbelt, Maryland, was widely known for his bold and innovative mission-design skills. Back

FIGURE *7.2*

Robert Farquhar, who headed the first modern study of a mission to Pluto, named *Pluto 350*. (Johns Hopkins University)

in the mid 1980s, when NASA couldn't get Congressional approval to fund a mission to Halley and was faced with the embarrassment of an international (and wholly non-U.S.) armada of spacecraft heading to the comet, the ever-innovative Farquhar discovered that one of NASA's high Earth-orbiting spacecraft could be sent from Earth orbit to a comet called Giacobini-Zinner. By using a then-untried slingshot trajectory past the Moon, Farquhar found that American spacecraft would arrive at comet Giacobini-Zinner 100 days before the onrushing armada was set to encounter Halley. Farquhar's plan was accepted, and it worked—making the United States the first nation to send a spacecraft past a comet.

But compared to simple comet flybys, Farquhar and the *Pluto 350* design team at JPL had a considerable challenge in front of them. As we have described, for 25 years NASA's planetary mission spacecraft had been steadily growing in size. By 1990, they typically weighed in at two to four tons. *Pluto 350* would have to reverse that trend.

After months of work, Farquhar and his team managed to pack communications, propulsion, pointing, and power systems into a compact structure weighing in at about 700 pounds, including the fuel needed to make trajectory corrections during the long flight to Pluto. The *Pluto 350* team also left room for a powerful suite of cameras, spectrometers, and magnetometers weighing about 35 kilos (50 pounds).

Farquhar, still the clever mission designer, realized that the cost-effective way to get *Pluto 350* to Pluto was to trade a large and expensive launch vehicle capable of shoving his spacecraft directly to Pluto, for a less expensive, medium-sized launch vehicle that need only lift *Pluto 350* to Jupiter. A close flyby of Jupiter could then be used to inject the vehicle on its Pluto trajectory, in just the same way that the *Pioneers* and *Voyagers* had used Jupiter to bootstrap themselves to Saturn.

But this was still not enough for Farquhar, a virtual Nureyev of orbital mechanics. Early in 1990, Bob Farquhar found that by launching toward *Venus*, and then making a series of Venus and Earth swingbys, *Pluto 350* could be catapulted toward Jupiter after launch from one of the smallest and least expensive launch vehicles—the *Delta*. Farquhar's solution could cost the Pluto mission time, but it would save a great deal of money that would otherwise be spent on expensive rocketry.

At the same time that Farquhar and the JPL design team were designing *Pluto 350* and its clever trajectory, Bagenal, Stern, and their colleagues in the Pluto Underground were charged with defining the scientific objectives and rationale for the mission. The researchers first summarized the central questions that could not be answered from ground-based or Earth orbital instruments, and which required on-site reconnaissance to resolve. In a report to the highest-level NASA advisory board directly responsible for planetary exploration, the Solar System Exploration Subcommittee (SSES), the Pluto Underground laid out their most burning questions including:

- *How did Pluto really compare with Triton?* Were they really close cousins left over from the early era when ice dwarfs were common, or were they only superficially alike?
- *Is Pluto's surface composition as variegated as its surface markings?* That is, are Pluto's surface ices uniformly distributed, with the same amounts of nitrogen, carbon monoxide, and methane at all locations, or is the nitrogen-dominated surface spotted with a checkerboard of frozen carbon monoxide and methane ponds?
- *How deep are Pluto's ices—that is, are they simply a thin veneer lying on the surface or are they instead piled on as a deep crust?* Are Pluto and Charon's dramatic compositional differences more than skin deep? Is Pluto internally active? How much is Pluto's geology like Triton's? (After all, despite the fact that the two planets have the same size and density, Earth's geology and Venus's are radically different.) How does Charon's geology compare with Pluto's?
- *What is the structure of Pluto's atmosphere?* How fast is it escaping to space? Are haze layers the cause of the kinked stellar-occultation lightcurve, or is the cause the atmosphere's thermal structure?
- *What are the effects of Pluto's strong seasonal changes?* Is this why Pluto's surface is so contrasty? Why are the northern and southern polar caps so different? Why is the darkest region on the planet directly under Charon, and why is the brightest region exactly 180 degrees away?
- *How did the Pluto–Charon binary come to be?* Was it indeed created through a giant impact, or was there something missing in our Earth-based view of Pluto, hinting that something else occurred?

The Underground's list of questions went on, and on, and on. There was much to ask of Pluto.

TICK, TICK, TICK

While the scientific community was formulating the specific questions it wanted answered by *Pluto 350,* Farquhar and his team worried that because Pluto had just passed perihelion, its increasing distance from the Sun would, with every passing year, make the trip time longer and the job of communicating with the spacecraft harder.

Almost simultaneously, the scientists found that time was working against them, as well. Why? After perihelion, as Pluto drew away from the Sun, it would start to cool. Because Pluto's atmospheric bulk depends so sensitively on the surface temperature, the best-available computer models predicted that the atmosphere would probably undergo a rapid demise in the decade between 2010 and 2020. This demise, which experts such as McDonald Observatory's Larry Trafton and one of his students had dubbed "atmospheric collapse," would result in a snow-out of most of the atmosphere onto the surface. If a reconnaissance spacecraft arrived after the collapse, there wouldn't be much (or possibly any significant) atmosphere to study.

To make matters worse, the onset of a deep southern-hemisphere winter (driven by Pluto's 118-degree polar tilt) would be plunging more and more of Pluto (and Charon) into a decades-long shadow, where it cannot be imaged or have its composition mapped. By 2015, almost half of the southern hemispheres of Pluto and Charon would be shrouded in darkness and thus hidden from view.

It wouldn't matter if a better or faster mission arrived in 2030, or 2050: As demonstrated in Figure 7.3, the opportunity to study Pluto's atmosphere and to globally map Pluto and Charon would be lost. The last time Pluto had been so favorably placed was in the middle 1700s; the next occurrence wouldn't come until the 2230s. With both the encouraging results of the *Pluto 350* study and a compelling scientific case for Pluto reconnaissance in hand, it was time to act.

CLIMBER'S ASCENT

Farquhar, Bagenal, and Stern took the case for a small Pluto mission to the upper echelons of NASA's space-science management. Simultaneously, they and the expanding roster of Pluto Undergrounders took the case up through the scientific review committee structure that NASA uses to debate and rank different mission proposals. In late February, 1991 (during the week of the Allied Gulf War ground campaign against Iraq), the SSES met to balance priorities for the 1990s.

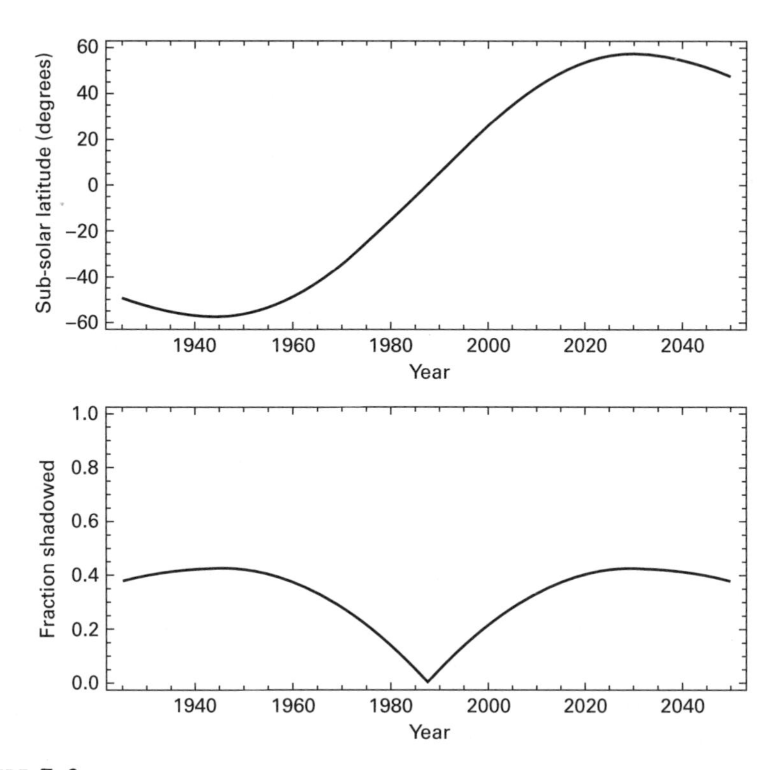

FIGURE *7.3*

As Pluto's aspect to the Sun gradually changes, the proportion of its surface perpetually in shadow also increases or decreases. The top graph shows the change in subsolar latitude (where the Sun appears overhead on Pluto) over time. The lower graph shows how the proportion of Pluto's surface experiencing perpetual night changes over the same period of time. Note that almost the whole of Pluto's disk was visible in the late 1980s when the subsolar latitude was zero. Since then, the fraction of Pluto's surface always in shadow has been steadily increasing. The longer spacecraft exploration of Pluto is delayed, the smaller the surface area that will be visible. By 2020, only 60% of the surface will be illuminated, and it will be many decades before the situation significantly improves again. (Courtesy Marc Buie)

The Pluto mission was popular among many members of the SSES, including its polished young chair, Jon Lunine of the University of Arizona. However, it was criticized by others, who said it had "come up through the ranks too quickly," or that it would be "too many years before the mission could return data." But Pluto's critics soon found themselves up against some formidable odds when Geoff Brigg's successor as NASA's new chief of solar-system exploration, Wesley

Huntress announced to SSES that the scientific proposal and public appeal of a Pluto mission led him to believe a Pluto mission should be one of his top three priorities. Still, more wind was taken out of detractor's sails when Don Hunten, one of the most respected figures in planetary science, stood up, looked around the room, and delivered a short but effective soliloquy. The 67-year-old senior scientist almost shouted: "I don't expect to be living when the Pluto mission arrives at Pluto, and if I am living, I don't expect to be aware of the event, but this is the kind of valuable research we should be doing. It's important. Let's get on with it." After two days of debate, the SSES wrote its report, which ranked a Pluto flyby in the highest priority category for the 1990s.

Less than two weeks later, NASA's Huntress formed an official science working group to guide detailed Pluto mission planning. This group was called the Outer Planets Science Working Group, or OPSWG. The group included more than 20 of the leading American experts on Pluto and the other outer planets. Among those selected for OPSWG were Fran Bagenal, Marc Buie, Dale Cruikshank, Jim Elliot, Bill McKinnon, Dave Tholen, Alan Stern, Larry Trafton, and Roger Yelle. OPSWG's chair, Alan Stern, and about half of OPSWG's members were straight from the Pluto Underground. The other half of the group's roster were *Voyager* and *Galileo* veterans. At OPSWG's first meeting, Wes Huntress (see Figure 7.4) remarked that the Pluto Underground wasn't very underground any more.

In the first months after OPSWG was formed, NASA asked the group to work with JPL to study a much larger mission than *Pluto 350*. That mission would have used a copy of the *Cassini* Saturn orbiter, weighing more than 3,800 kilograms (8,000 pounds). Although it could have carried a whole host of sophisticated instruments and a daughter spacecraft or atmospheric probe to deepen the investigation of Pluto, the cost of this mission concept was estimated to be over $2 billion. Given the ever-tighter NASA budgets and the consequently declining resources for planetary exploration, a lead balloon could not have sunk faster. OPSWG recommended that NASA opt for the lighter and less expensive *Pluto 350* concept. For more than 20 years, NASA science working groups had designed increasingly complex missions with greater and greater instrument rosters and, by the way, increasingly greater price tags. OPSWG's decision to recommend *Pluto 350* was the first time anyone could recall that a scientific committee of NASA had opted for a smaller mission than NASA itself had recommended. What OPSWG didn't know, however, was that a far smaller and still more radical mission concept was cooking behind the scenes at JPL.

FIGURE *7.4*

Left to right—Alan Stern (OPSWG's chairman), Wesley Huntress (Director of NASA's Office of Space Science), and Richard Terrile (JPL Pluto mission scientist) at a meeting of the Outer Planets Science Working Group (OPSWG) in 1993: The OPSWG was formed by NASA in March, 1991, but disbanded in mid-1996.

Enter Rob Staehle and Stacy Weinstein (see Figure 7.5). Staehle, a savvy systems engineer, and Weinstein, an expert trajectory designer, were about as fired up as any Underground member had ever been. Back in 1989, they had seen a new stamp set, issued by the U.S. Post Office, that featured the highlights of planetary exploration. For Venus, there was *Mariner 2.* For Mercury, a stamp showcased *Mariner 10*; a Mars stamp glorified the *Viking* landers. The Jupiter and Saturn stamps featured the trailblazing *Pioneer 10* and *11* spacecrafts, and the Uranus and Neptune stamps focused on *Voyager.* The Pluto stamp (Figure 7.6) said, simply: "Not yet explored."

Staehle and Weinstein took the Pluto stamp as a call to arms. They didn't know much about Pluto, or its scientific value at the time, but they damn-well had a fire in their bellies to explore this last frontier!

As it turned out, Staehle and Weinstein had never heard of the Pluto Underground—it may be a small world, but engineers and scientists run in different circles. Thinking they were alone in the world in advocating a Pluto mission, Staehle, Weinstein, and a maverick spacecraft designer named Ross Jones spent 1991 begging, borrowing, and virtually stealing

FIGURE *7.5*

Robert Staehle and Stacy Weinstein, pioneers of the *Pluto Fast Flyby* concept. (Alan Staehle)

FIGURE *7.6*

The stamp for Pluto in a 1991 U.S. set devoted to the exploration of the planets by spacecraft alone declared "*Not Yet Explored.*" It represents a challenge and a rallying point for Pluto enthusiasts determined to make a Pluto mission a reality.

JPL resources to engineer a truly tiny but capable Pluto reconnaissance mission they called *Pluto Fast Flyby,* or *PFF* (see Figure 7.7).

At 140 kilograms, including both onboard fuel and its scientific instruments, *PFF* would weigh less than half what the already-small *Pluto 350* would! In fact, *PFF* would be so light that it could be launched by a large launch vehicle on a fast trajectory directly to Pluto (see Figure 7.8), thereby avoiding the elegant but time-consuming Venus, Earth, and Jupiter swingbys that Farquhar had designed. As a result, *PFF* could make the 5-billion-kilometer (3-billion-mile) crossing in as little as seven or eight years, at an average speed almost twice what *Pluto 350* would achieve. By reducing spacecraft size, mass, and complexity, *PFF* would be significantly cheaper than *Pluto 350.* By reducing flight time, *PFF* would also save operations costs and allow designers to relax the stringent subsystem reliability requirements that a long, 13-to-15-year mission such as *Pluto 350* would impose. It was a very attractive package, but it had one catch.

The main drawback of the *PFF* concept was that the spacecraft couldn't carry a very large scientific package. After all, the package of

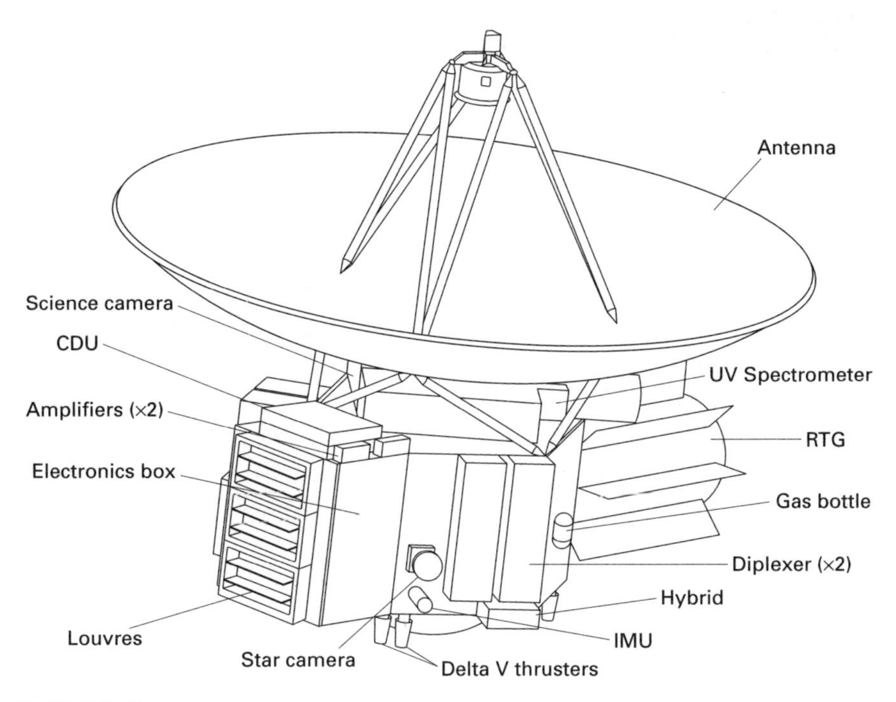

FIGURE *7.7*

An early version of the *Pluto Fast Flyby* craft. (JPL)

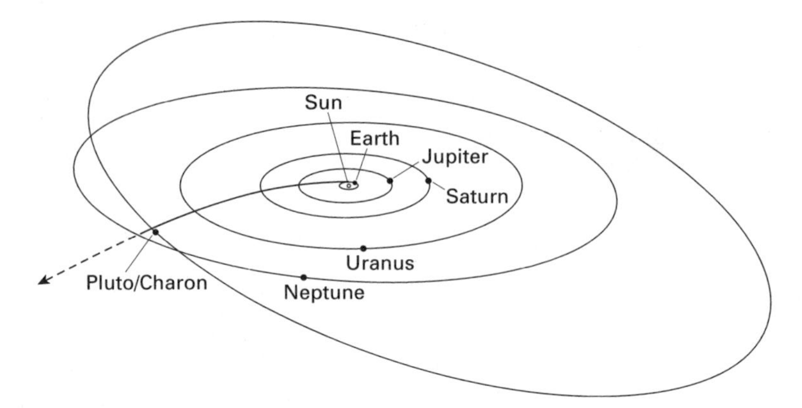

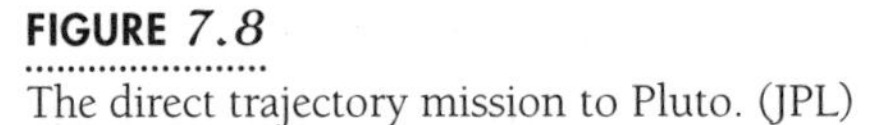
FIGURE 7.8
The direct trajectory mission to Pluto. (JPL)

scientific instruments on recent NASA missions such as *Galileo* and *Cassini* had weighed more than the entire *PFF* spacecraft! Staehle's team initially estimated that *PFF* might be able to carry five kilograms (12 pounds) of instruments. The single lightest instrument on *Cassini* weighed almost twice that.

OPSWG swallowed hard. The *Pluto 350* call had been difficult for many of the scientists on OPSWG, but *PFF* was the equivalent of taking one's first bungee jump. "Will this thing really work?"

It was not an easy decision, and OPSWG debated the trade-off between the two missions during the first half of 1992. By summer, the pragmatists on the panel prevailed. They lobbied that a focused mission with three or four miniaturized sensors was sufficient to perform the basic reconnaissance needed at Pluto, including mapping, and that given the declining budget at NASA, it was unlikely that the larger mission, *Pluto 350,* would ever get off the drawing boards. The vote OPSWG took was a landmark event, with a 3:1 majority favoring *PFF*.

Now, however, OPSWG had to prove to their colleagues on the SSES and in the larger planetary-science community that a mission as light as *PFF* was worth flying, from a scientific standpoint. To do this, OPSWG outlined a focused set of reconnaissance objectives that it argued constituted a bottom line, below which it wasn't worth going to Pluto. They weren't fooling around—this list of core objectives, which they dubbed Category 1A, was spare indeed. It consisted of just three items, which were ranked equally. After all the scientific jargon is removed, this is what OPSWG's Category 1A objectives said *PFF* had to accomplish if the mission was to be worth flying:

- Characterize the global geology and geomorphology of Pluto and Charon by mapping each body in its entirety at a resolution of 1 kilometer (about 0.6 miles) or better.
- Obtain high-spectral-resolution infrared images of the surfaces of both Pluto and Charon, at a resolution of 10 kilometers (6 miles) or better in order to map their surface compositions.
- Accurately determine the composition and structure of Pluto's atmosphere, and characterize the details of any haze layers in Pluto's atmosphere.

There wasn't any fat in this list. OPSWG—Plutophiles and all—recommended that if *any one* of these objectives couldn't be achieved, *PFF* should be abandoned. Although the OPSWG's three *PFF* objectives created a stable scientific base from which to form a solid, first-order view of the Pluto–Charon system. Leaving any one of the objectives out (mapping, composition, or atmospheric assay) would result in a deeply incomplete characterization that simply wasn't worth the effort.

Despite this, *PFF* was too big a step for many, and it came under heavy outside attack. Within the larger scientific community, many researchers said OPSWG had swayed too far under pressure by choosing a concept too bold in its engineering targets and too constricted in its scientific breadth. Additionally, within NASA and JPL, some bureaucrats and program managers assailed *PFF* for its untested capabilities.

Seeing that his baby was in trouble, Staehle went straight to the top. Staehle learned that NASA's chief, Daniel Goldin, was going to be near JPL a few weeks later to present a motion picture Oscar that had flown on the Space Shuttle at a small ceremony of the Motion Picture Academy of Arts and Sciences (see Figure 7.9).

Staehle went to the ceremony, introduced himself to Goldin, and said that his group of young engineers had a scientifically endorsed plan to get to Pluto with a revolutionary, small spacecraft, but that The Establishment wouldn't let the plan reach the light of day.

Administrator Goldin then was at the very outset of his campaign to reinvent NASA, and Staehle knew it. Goldin listened to Staehle and quickly decided to make both Staehle and Pluto exploration a symbol of his New NASA. Staehle's chutzpah had paid off—Goldin went back to Washington and annointed the Pluto mission as his own. Staehle's bold maneuver peeved *PFF*'s detractors, but the quantum leap he'd made by enlisting Goldin directly pushed *PFF* into the spotlight, and toward the center of NASA's planning for the 1990s.

FIGURE *7.9*
Dan Goldin, head of NASA since 1992. (NASA)

FORM AND FUNCTION COMBINED

When Dan Goldin emblazoned the badge of Pluto exploration on his sleeve, he took on a special responsibility, for the Pluto mission represented something particularly distinctive in all that was and is the U.S. space program.

It was, as we have said, the Everest of exploration. But it was also something deeper: For many, the idea of a mission to Pluto, at the very frontier of our solar system, rekindled sthe spirit of hope and unbridled horizons that they remembered from the explorations of *Apollo*.

With Dan Goldin's support, the drive to send a mission to Pluto moved forward rapidly in the mid-1990s. Goldin poured in funding to overcome some of the mission's most daunting challenges. Within the space of two years, the resources Goldin devoted toward *PFF* yielded miniaturized thrusters the size of a nickel, new-tech frequency communications gear, and a dozen other nitty-gritty technical improvements needed to make *PFF* feasible.

Perhaps the most revolutionary advances of all were those in the area of science instrumentation. Based on OPSWG's 1A objectives, *PFF* would have to pack a sophisticated camera, plus an ultraviolet spectrometer and an imaging infrared spectrometer, into a spare, 7-kilogram (16-pound) package.[24]

Many in the scientific community thought this was a near impossibility, and they had good reason to be skeptical. *PFF* would have to reduce into just seven kilograms the capabilities of a set of high-tech *Cassini* Saturn-orbiter instruments with a mass of 108 kg. Further, whereas the *Cassini* cameras and spectrometers consumed more than 90 watts and cost almost $100 million to produce, the Pluto package would have to consume less than 5 watts and cost no more than $30 million—for two copies! It was equivalent to packing fighter plane performance into a machine the size of an eagle, and doing it for two bits on the dollar.

To prove that this could be done, OPSWG and Staehle's team asked NASA to fund engineering demonstrations of *PFF* instruments. So in February of 1993, NASA announced its intention to select several proof-of-concept Pluto instruments for advanced development. Less than 90 days later, the space agency whittled down the proposals submitted by industrial, university, and government labs to a handful of winners, which it funded on rapid-prototyping, 12-month schedules to create working models of each micro-instrument.

NASA funded two main approaches to miniaturizing the Pluto payload. First, they would fund a set of miniaturized cameras and spectrometers that could be flown as a set to cover OPSWG's 1A objectives. Second, NASA would also fund a more revolutionary approach: combined instrument packages that would meld together the whole experiment suite into one sleek, ultra-efficient unit.

The so-called *integrated* instruments combined the two spectrometers and the Pluto camera into a single unit, sharing structural and electronic resources; some even shared optics, channeling the precious light into parallel channels so that one or two telescopes would feed all three detectors. The various integrated-instrument designs also experimented with exotic materials, advanced manufacturing processes, and compact sensor heads.

---

[24]Two more kilograms were allotted to a radio experiment, which would probe Pluto's lower atmosphere.

The most radical aspect of the integrated instruments developed, however, was what one such instrument, the highly integrated Pluto payload system, or HIPPS (see Figure 7.10), called a "seamless boundary between spacecraft and instrument, . . . a payload without walls." A talented rival team to HIPPS, building an integrated instrument prototype called PICS (Pluto Integrated Camera System) adopted the same philosophy, which they dubbed the *sciencecraft* approach.

What the sciencecraft philosophy centers on is the revolutionary notion that *PFF* and its instrument set should be sewn so tightly together in design and function that the two would become one, saving mass, power, and dollars in the process. This holistic approach to engineering stood the traditional set of spacecraft black boxes, each with their own separate walls, electronics, and functions, on its head. If the sciencecraft approach could be made to work, the engineers' design job

FIGURE *7.10*

The highly integrated Pluto payload system (HIPPS) was designed to conduct a full reconnaissance of the Pluto–Charon system while being as compact as possible and keeping the weight of the spacecraft to a minimum. Compare the size of the HIPPS ultraviolet spectrograph with the human hands holding it!

would be tougher, but the resulting spacecraft would be leaner, sleeker, and closer to Staehle's Olympian ideal.

The stringent payload mass and power targets for *PFF* were tough to reach. By mid-1994, five of the seven selected instrument-development teams had succeeded (and two had failed) but paper concepts had in fact been turned into working laboratory models that met all of the performance criteria necessary to meet OPSWG's 1A science objectives.

Of the five teams that produced working prototype instruments within PFF's mass and power targets, PICS and HIPPS seemed to reign. Why? Because these integrated-instrument packages were designed to blend seamlessly into the *PFF* spacecraft, thereby producing additional cost and mass savings. In fact, with either HIPPS or PICS, the Pluto mission could expect to return more information, more detailed results, and a more well-rounded view of Pluto than *Voyager*'s heavier and more complex suite of 11 instruments could have produced.

Later in 1994, NASA's Goldin asked for an independent review panel chaired by military spacecraft designers skilled in miniaturized space missions to assess the feasibility of *PFF* and its miniaturized instruments. The review panel conducted an in-depth, month-long technical review and reported to Goldin a resounding endorsement of *PFF*. In less than 18 months, *PFF* had vaulted from little more than a daring concept to a startlingly feasible set of prototypes. Staehle and his JPL team, as well as the OPSWG scientists, felt ready to be turned loose to build and launch *PFF*. But this was not to be.

SOME TROUBLE WITH THE TICKET

As envisioned since its inception, *PFF* would involve two spacecraft launched a few weeks apart. OPSWG had demanded this, in part to mitigate the risk of a single-spacecraft mission, and in part because Pluto–Charon rotates on its axis so slowly that two spacecraft would be needed to map both sides.[25]

However, for *PFF* to reach Pluto quickly—one of its primary goals—*PFF* would have to be more than just light; it would also have to be launched aboard one of the two most powerful launch vehicles in the U.S. inventory: The *Titan* or a Shuttle. *Titan* was far preferred by

[25]If one spacecraft were sent out, then Pluto's 6.4-day rotation period would allow it to visit only one side of Pluto close up. The other side would last be seen 3.2 days—half a rotation—before closest approach. At that point, millions of kilometers out, far fewer details would be seen.

most of Staehle's design team and by most of OPSWG because it is often far less expensive to integrate an unmanned spacecraft with an unmanned rocket, than with the manned Shuttle flight, where safety reigns.

Still, the U.S. Air Force had a monopoly on *Titan*'s services and wanted to charge NASA almost $500 million for each launch. At that price, *PFF*'s two spacecraft would cost almost a billion dollars to launch—almost twice the total cost of two *PFF* spacecraft, their instruments, and flight-operations costs. To Dan Goldin, this represented an untenable balance. The cost of *Titan* had become a Pluto killer, and something had to be done about it.

The Plutophiles met this problem with a double-pronged attack. At JPL, mission designer Weinstein found a series of Farquharesque trajectories that used successive Venus and Earth flybys to swing *PFF* to Jupiter, and then on to Pluto. These trajectories would add three years to the trip,[26] but *PFF* could then be launched using any one of several smaller and far less costly launch vehicles than *Titan*.

At the same time, OPSWG's chair, Stern, set out to remove even more of the launch costs from NASA's books by internationalizing the mission. This in mind, he set off an unofficial trip to Moscow in January of 1994, in hopes of making inexpensive Russian launchers—perhaps, he hoped, even paid for by the Russian Space Agency—another option.

In Russia's capital, Stern gave briefings on the Pluto mission to more than 50 leading space scientists and program managers at the main astronomical center of Moscow's largest university—Moscow State's Shternberg Institute—and at Russia's premier center for scientific space exploration—the Institute for Cosmic Research (the Russian acronym for which, IKI, is pronounced "Ee-Kee").

The pivotal moment of the trip came after the briefings, on a gray winter's afternoon in IKI Director Alec Galeev's office (see Figure 7.11). Stern wanted to know from Galeev ("Gal-ye-ef") whether Russia might be interested in contributing one of its powerful launch vehicles in exchange for Russian participation in the mission. Galeev's response was, simply, "nyet," because he didn't see much in NASA's Pluto mission for Russia, or for Russian scientists.

Stern wondered what the Russians might consider a useful "piece of the pie" and recalled the one real aspect of the old *Mariner Mark II* Pluto-mission concept. Stern asked Galeev whether IKI and the Russian

---

[26]Staehle combatted the added mission operations costs by finding ways to slim down the necessary operations teams to a few dozen people.

FIGURE *7.11*

Alec Galeev, Director of the Russian Institute for Cosmic Research, IKI, in Moscow. (A. Stern)

space-science community might be more interested if *PFF* carried a Russian-built probe designed to enter Pluto's atmosphere. Galeev sat up in his chair, smiled, and said, "*This* is an interesting idea." To Galeev, the idea of a Russian Pluto entry probe offered many attractions. First, there would be the highly visible nature of the brief but daring probe mission. Second, it would give the Russian space science community a unique and valuable data set to call their own. But most importantly to Galeev, this might be the way to move Russia's planetary exploration program into the outer solar system for the first time, "perhaps by sending key engineers to the U.S. to learn what NASA had mastered with the *Pioneers* and *Voyagers*."

EXPRESS

Within weeks, the seed that Galeev and Stern planted in Moscow took root. When Stern brought the concept home to NASA and JPL, Staehle and his team were excited about the prospects for a joint mission; so was Wes Huntress, who had by then been promoted to the position of being NASA chief of all space science. Huntress briefed his boss Dan Goldin on the concept and found him enthusiastic.

Within a few more weeks, tentative letters were exchanged between Moscow and Washington, suggesting that *official* exploration of a joint mission take place. Over the next few months, U.S. and Russian teams exchanged data and developed a technical plan for the joint-mission concept. The plan called for the two U.S. Pluto spacecraft to be launched by either the large Russian *Proton* or the somewhat smaller and less expensive Russian *Molniya* rocket. Each of the two U.S. *PFF* spacecraft was also to carry Russian probes called, "Drop Zonds"[27] which were to dive into Pluto's atmosphere as the tiny mothership hurtled overhead.

At the same time that U.S.-Russian cooperation plans were being sketched out, NASA asked Stern and JPL's Pluto mission scientist, Richard Terrile (see Figure 7.4), to make a 10-day round-robin tour of European space centers, describing the Pluto mission and investigating whether other potential partners might be interested in joining in *PFF* as well. The goal of these meetings was to further internationalize the mission in order to reduce costs. Terrile and Stern were joined for part of the trip by space physicist and OPSWG member David Young. The three scientists gave briefings on Pluto mission planning and scientific objectives to German, Dutch, Swedish, French, Italian, and Spanish scientists.

DETOUR

By the end of 1994, the NASA and JPL were planning on Russian cooperation, and conducting a detailed study of a German proposal to provide a Drop Zond-like probe to study Io's atmosphere during the planned Jupiter flyby as well. Equally important, working models of the advanced, miniaturized instruments and other crucial hardware components were being delivered to JPL.

PFF's main focus was moving from feasibility studies to tactical planning for what NASA officially refers to as a "new start" when it solicits approval and funding to begin building a space mission.

However, just as that shift in focus was beginning, NASA's Goldin mandated still further reductions in the mission's size and cost. Goldin wanted to see a lighter spacecraft that would showcase even-more advanced technologies.

Despite this being their third redesign for Goldin, Staehle and his JPL team responded positively by re-examining every gram of structure for savings, and by using, wherever they could, advanced communications, power generation, and propulsion components barely off laboratory testbeds. Staehle also enlisted the help of a University of Colorado spacecraft operations team led by Elaine Hansen. Hansen was famous for her innovative, low-cost approach to mission operations.

---

[27] *Zond* is the Russian term for a space probe.

With Hansen's help, Staehle and his team found that they could use high levels of computer automation to operate Staehle's elegantly simple Pluto mission with as few as ten people at JPL, including trajectory designers, flight controllers, and the mission chief. By contrast, missions such as *Voyager, Galileo,* and *Cassini* had been staffed by hundreds (and at their peak, almost a thousand) of JPL flight-operations personnel.

Working hard through 1995, Staehle's team blended together more than 20 advanced technologies, a new mission-operations concept, and a sparing mission-management concept that would set a new standard for deep-space missions. When he and his team were done, they had whittled the spacecraft mass down to 75 kilograms—without affecting the valuable payload mass—and the spacecraft's power requirements to just 75 watts—about that of a single household lightbulb. They also chose a new name—*Pluto Express.*

Staehle's detailed cost estimates pegged the cost to develop both of the *Pluto Express* spacecraft (see Figure 7.12) and their instruments,

FIGURE 7.12

An artist's impression of the *Pluto Express* sciencecraft, with Pluto and Charon in the background. (JPL)

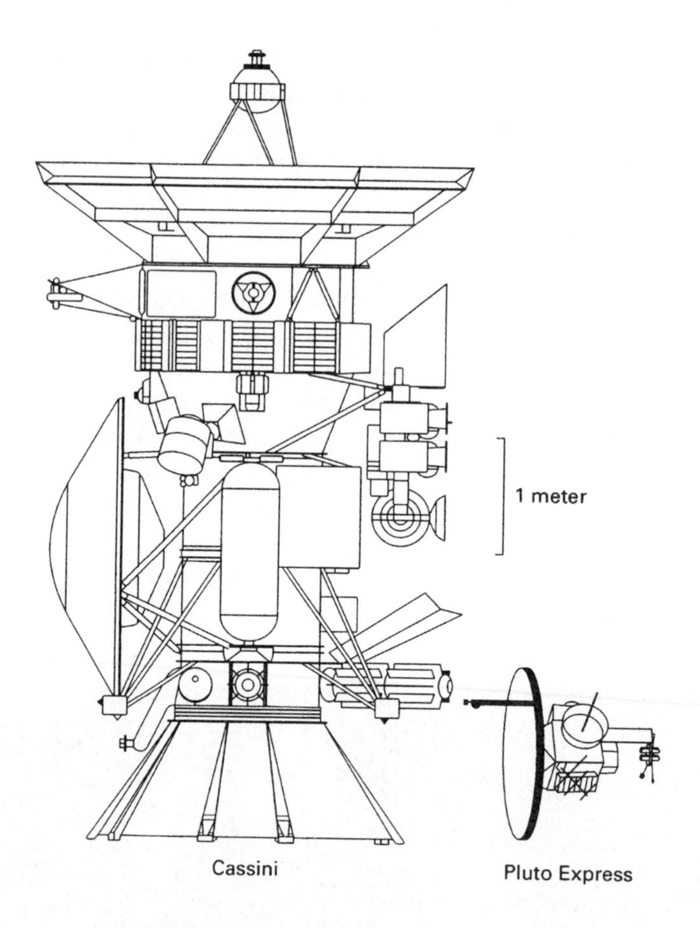

FIGURE *7.13*

A size comparison between the *Cassini* spacecraft (for exploring the Saturnian system) and the proposed *Pluto Express* mission. (JPL)

plus all of the necessary launch preparations and the flight operations system, at just under $350 million. Adjusting for inflation, that's about one-fourth the cost of *Voyager 2* alone. Many NASA Earth-orbiter missions cost more to build than the Pluto mission would! All told, Staehle could develop, build, and test the two Pluto spacecraft for a little over a dollar per American.

If NASA could muster the funding, *Pluto Express* would launch between 2001 and 2003. Each *Pluto Express* would then use multiple Farquhar/Weinstein-style swingbys of Venus or Earth (depending on the exact launch date) to rev up its energy enough to reach the Jupiter swingby. See Figures 7.13 and 7.14. At Jupiter, each *Pluto Express* spacecraft would concentrate on the volcanic moon Io, with its Ger-

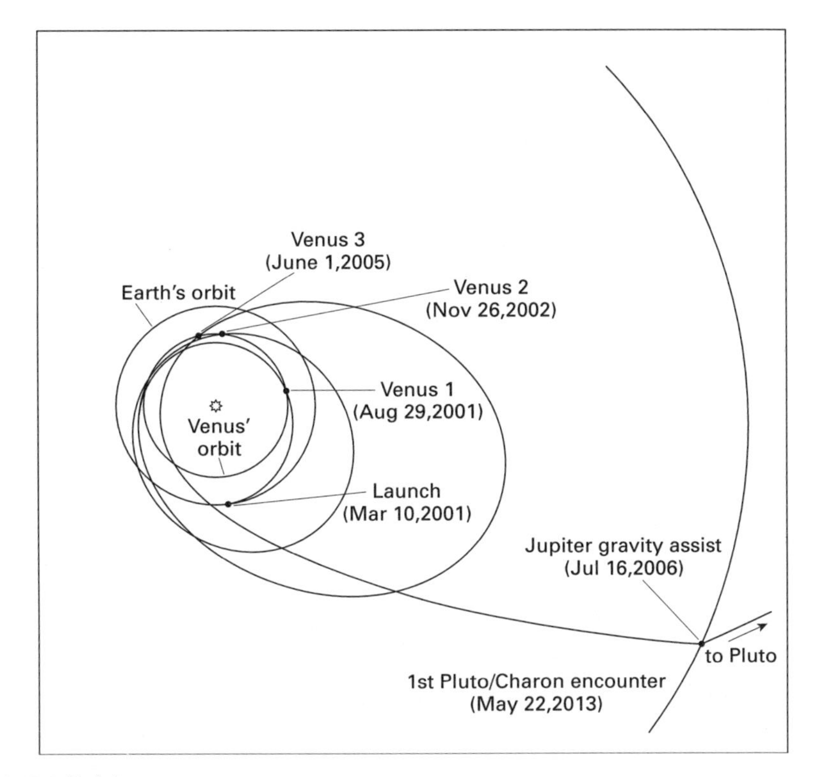

FIGURE *7.14*

The proposed trajectory for a *Pluto Express* (example launch shown for a March 10, 2001 launch). This trajectory relies on three close encounters with Venus, and one with Jupiter, to accelerate the craft on its way for encounter with Pluto, taking advantage of the gravity-assist technique.

man entry probe serving as the focus of the scientific investigations there.

The close pass each *Pluto Express* would make to Jupiter would then catapult it forward on a nearly beeline trajectory to Pluto. With an average speed near 65,000 kilometers per hour (40,000 miles per hour), the two spacecraft would need about 8.5 years to cross the gulf of the outer solar system.

The first and second Pluto flybys would be timed to be about six months apart, so that scientists could optimize the details of the second encounter, based on what was found by the first (Figure 7.15). One spacecraft would concentrate on studying one side of Pluto; the other spacecraft would concentrate on the opposite side of the planet. Each flyby would also survey Charon, and each would aim a Russian Drop Zond into Pluto's atmosphere. Each spacecraft would measure the com-

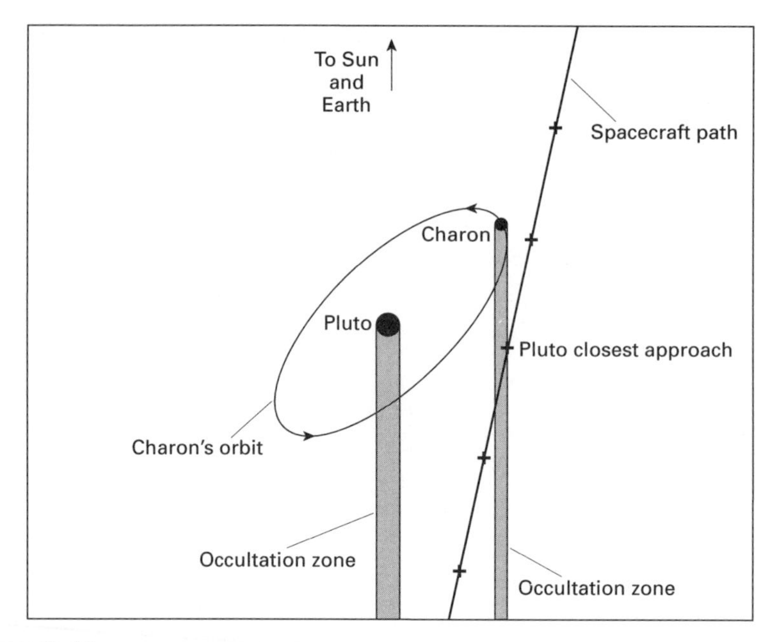

FIGURE *7.15*

The path *Pluto Express* would take during its flyby encounter with Pluto and Charon. The crosses mark intervals of 20 minutes.

FIGURE *7.16*

This *Voyager* 2 image of complex terrain on Uranus's moon Miranda gives a good idea of the level of detail *Pluto Express* would reveal about the surface of Pluto. The resolution here is about 800 meters (2,600 feet). The large impact crater at lower center is about 25 kilometers (15 miles) across. (NASA/JPL)

position and structure of Pluto's atmosphere and would map the surface geology and composition of the side of Pluto (and Charon) it would fly by. Additionally, each spacecraft would search for small satellites in the Pluto–Charon system, refine mass and density measurements of Pluto and Charon, and actually determine the rate at which Pluto's atmosphere is escaping into space.

The Russian Drop Zonds would carry a mass spectrometer deep into Pluto's atmosphere to bring a new dimension to measurements of the atmosphere's chemical composition, and to make measurements to determine whether Pluto has a magnetic field.

If all went well after passing Pluto, one or both *Pluto Express* spacecraft could then be sent on to make flyby mapping and composition studies of one of the myriad Kuiper Belt mini-planets that orbit in the frigid wilderness beyond the ninth planet.

HILLARY'S PAUSE

It has now been more than 8 years since the Pluto Underground first convened in a smoky Italian diner in Baltimore, and more than 6 years since NASA moved Pluto exploration to its "front burner." That's a very long time, and much has changed as the mission has evolved through several major incarnations. In an era of declining NASA funding, it is a testament to the attraction of Pluto that the mission continues to chug along, very slowly, but very surely all the same—the little engine that could.

Pluto Express is shaping up to be a highly ambitious scientific-exploration mission. If approved for flight, it would provide the first-ever probes onto a Jovian satellite, twin reconnaissance flybys of the Pluto–Charon system, and then a daring extended mission into the ancient realm of the Kuiper Belt. This mini-grand tour of the deep outer solar system would be accomplished at a cost that's about one fourth that of the *Voyager, Galileo,* and *Cassini* missions that were its predecessors.

The rub, of course, is whether the project will ever move from study to development. When the Pluto Underground came together and began their quest, they created and often used the climbing of Everest—Earth's highest peak—as a metaphor for the exploration of Pluto by spacecraft. What they did not realize is that there are two Everests that the Pluto mission must scale. There is, of course, the actual journey to the Everest that is Pluto, but there is also the equally difficult journey to ascend the Everest of NASA and Washington politics. Maddeningly, it looks as if it will take NASA more years to get the

Pluto mission out of the Washington, D.C. beltway, than *Pluto Express* would need to cross the whole solar system!

On the Eartlhy Everest of Washington politics, *Pluto 350,* a.k.a. *Pluto Fast Flyby,* a.k.a. *Pluto Express,* has reached the equivalent of Sir Edmund Hillary's last camp before his final assault on the mountain's stratospheric summit.

We do not know whether the *Pluto Express*'s final assault toward approval will be successful, but we do know the Plutophiles will not retreat without reaching the summit.

CHAPTER

8

POSTSCRIPT: WHERE NO ONE HAS GONE BEFORE

Ad Astra per Aspera (To the stars through difficulties).
KANSAS STATE MOTTO

NINTH SISTER

For 4 billion years, Pluto and Charon orbited the Sun in cold and quiet anonymity, but in the last fraction of the last percent of that span came something new. That something happened on Sol's third planet, and it was the rise of *our* species—the first in Sol's system to cast its gaze and intellect to the heavens. And when we *Homo sapiens* did just that, we found a universe matching our greatest hopes and aspirations. So, too, we found what at first certainly seemed only a footnote to our universe and our planetary system—Pluto.

Somehow, the last among the known planets took on a special attraction, apparently beyond all proportion to her mass and status among her comparatively gargantuan sisters. Perhaps it was just the mystery of this little world so far away, or perhaps it was Pluto's diminutive size and underdog position among the planets that made her, in many ways, what the Russians call "nasha dyevoshka," our dead little girl.

Whatever it was that sparked the palpable attraction between people on the third planet and the ninth sister of the solar system, it generated a torrent of discoveries, and Pluto revealed herself to be no footnote at all. What a wonderful thing we found Nature had concocted out there from a few ices and minerals, a little classical low-temperature physics, and a time span a thousand times our own!

Still, after 60 years of study, so many of Pluto's secrets remain, guarded by the protection that its distance and diminutive size so uniquely impose. We, as astronomers, the heaven-looking ambassadors of our inquisitive species, try to tear those secrets open and to riddle them with light.

What drives this thing in just one species, ours, among the myriad life forms that Earth has produced in four eons? Why do we, the people of Earth, need to explore? We do not know the basis behind this drive, but we do know it makes us feel alive!

AN ORNAMENT FAR

As we probe the planets and the stars and the galaxies with all our wit, and the newest tools of our technology, we often succeed—though the road of exploration is rarely a straight one. Sometimes we even fail, but the best times are when a new, hard-won fact reveals a dozen even newer questions beneath it.

Just such a day came on March 8, 1996, for this was the day that the world saw the true face of Pluto for the first time. It was the best of our long eyes that did it—the Hubble Space Telescope. Thanks to its perch, high above Earth's shimmering atmosphere, the looking glass of this fantastic telescope split the little spot that had always been Pluto in our telescopes into a disk and a mottled surface. Hubble had managed something no other telescope could. Recall that Pluto's diameter is just 0.1 arcseconds (1/36,000 of a degree): Imaging it is as difficult as counting the spots on a soccer ball 600 km (375 miles) away!

The Hubble images are crude (Figures 8.1 and 8.2), and faint, and fuzzy, and they are tantalizing. They reveal that Pluto is indeed an amazingly complex object, with more intricate large-scale surface patterns than any other world we know, except Earth and Mars.

Pluto's surface displays a dozen distinctive provinces, some of which are more than 1,000 kilometers (600 miles) across. These include a ragged northern polar cap, clusters of bright and dark spots in Pluto's equatorial regions, and a bright linear marking that *might* just be a long ejecta ray from an unusually large crater. The Hubble images and the map produced from them proved that Pluto's surface has the most complex network of features seen at this resolution anywhere in the outer solar system. The Hubble images and maps constituted a wonderful achievement, but also a melancholy one, for they are the best that any existing or planned telescope can do from our perch, 5 billion kilometers from Pluto.

Today, most of Pluto's long-held secrets remain secrets—guarded by

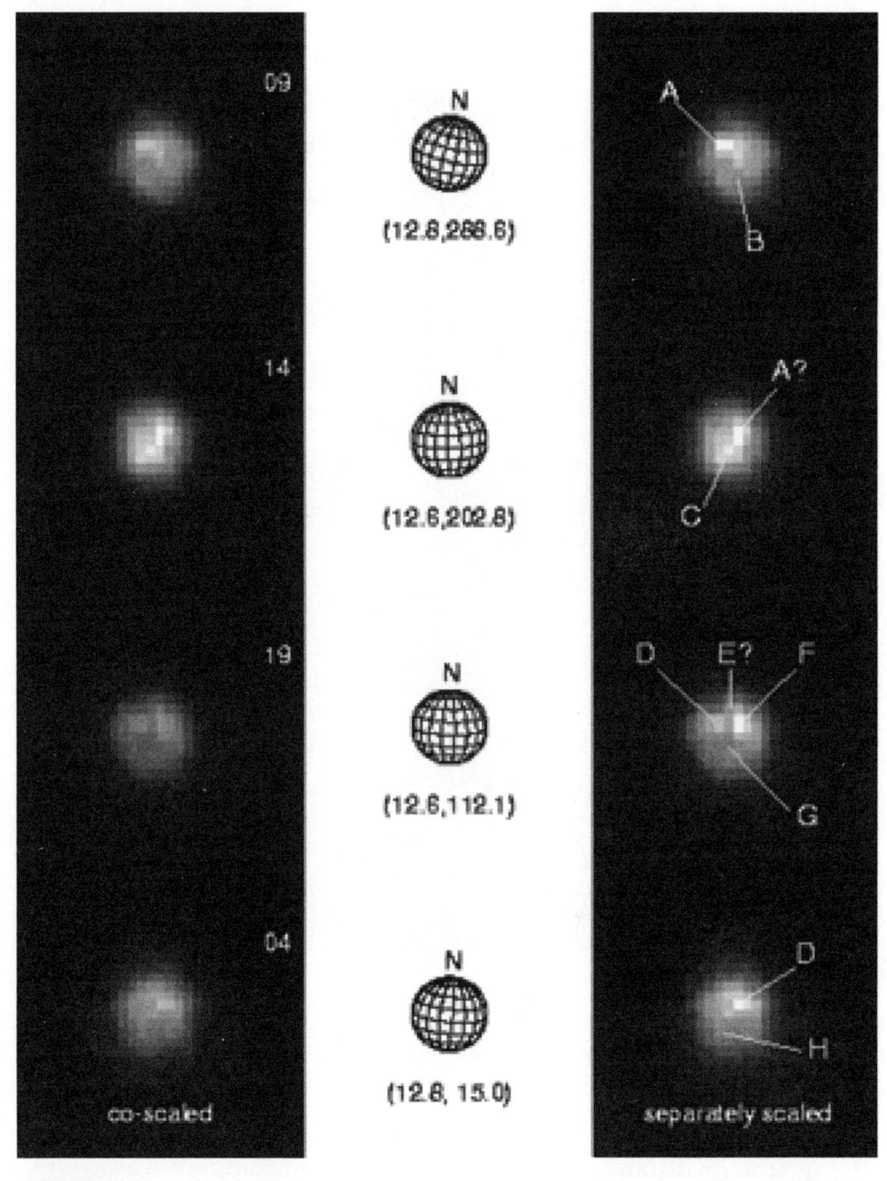

FIGURE *8.1*

Images of Pluto obtained with the Hubble Space Telescope (HST) in 1994, and first released after detailed image processing in March of 1996: Pluto's north pole is at the top in all images. Rotation advances from top to bottom by approximately a quarter of a turn between frames. The brightness of the images in the left column is scaled so as to preserve the relative intensity between frames. The images in the right column are scaled individually to show the maximum detail. The letters A through H label prominent features in the images. Note that some features are detected in more than one image as Pluto rotates (e.g., D). The center column shows a wire-frame grid scaled to the apparent size and geometry of Pluto for each image. Below each are given the latitude and east longitude for the sub-Earth point at the time of midexposure. (S. A. Stern and M. W. Buie)

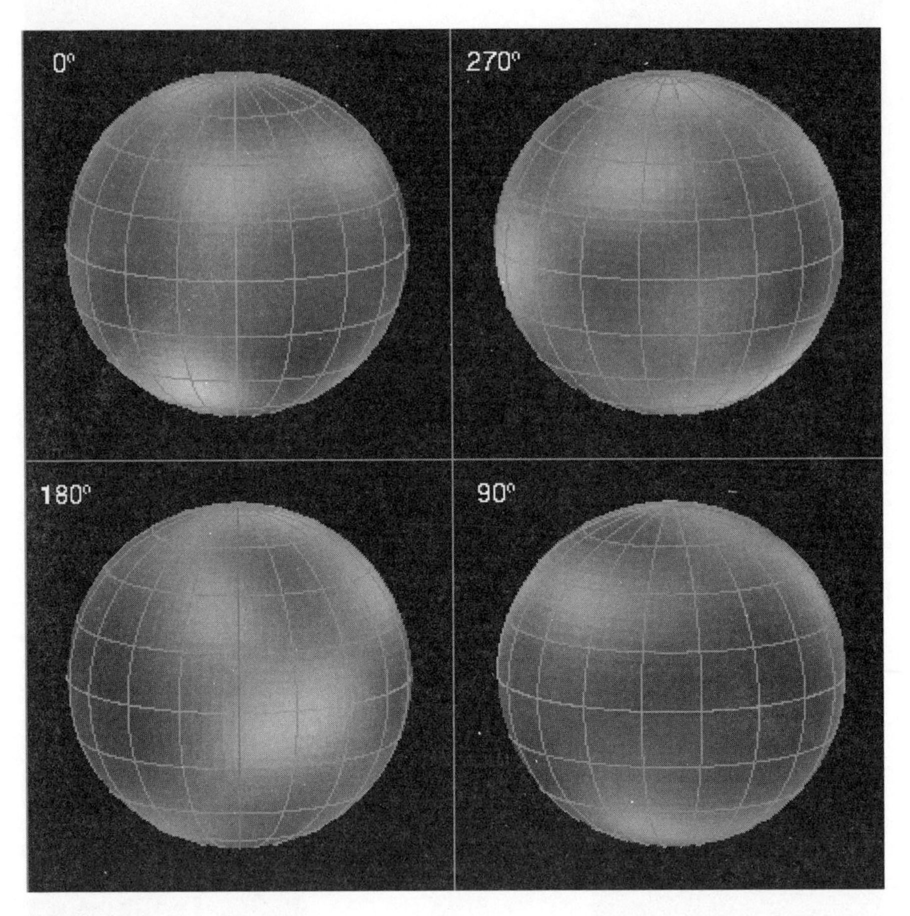

FIGURE *8.2*

The map of Pluto inferred from 1994 Hubble Space Telescope images projected onto four globes, differing by a quarter of a turn. (S. A. Stern and M. W. Buie)

the depths of distance between us and that ancient, icy little world. Hubble delivered a glimpse of the pinkish little ice-ball ornament to us, temptingly close now, but still too maddeningly far.

To see the solar system's ninth sister as she really is, we must go to her. And amazingly, our species has developed the will, and the way, to do just that.

So guard your secrets while you can, Pluto! We are coming.

A CHRONOLOGY OF MAJOR EVENTS IN THE EXPLORATION OF PLUTO & CHARON

- 1905 Percival Lowell undertakes his search for Planet X beyond Neptune. He will not succeed.
- 1916 Percival Lowell dies.
- 1929 Under the direction of Vesto Slipher, Lowell Observatory hires Clyde Tombaugh to carry out a new, systematic search for Lowell's trans-Neptunian planet.
- 1930 Pluto is discovered by Clyde Tombaugh on February 18. The discovery is announced publicly on March 12–13.
- 1930 Pluto's orbit is determined.
- 1955 Pluto's 6.4-day rotation period is determined (Walker & Hardie).
- 1965 Pluto's 2:3 orbital resonance with Neptune is discovered (Hubbard & Cohen).
- 1970 First crude spectrum of Pluto is published; Pluto's red color is firmly established (Fix et al.).
- 1976 Methane ice is detected on Pluto's surface (Cruikshank, Pilcher & Morrison).
- 1978 Charon is discovered, and from that, the total mass and average density of Pluto and Charon are determined (Christy & Harrington).
- 1978 Pluto–Charon mutual events are predicted (Andersson).
- 1980 Stellar occultation reveals Pluto's radius is just over 600 kilometers (Walker).
- 1983 IRAS satellite makes infrared measurements of Pluto's surface temperature.

- 1985 Onset of mutual-occultation events between Pluto and Charon (first unambiguous detections by Binzel, and shortly thereafter, Tholen).
- 1986 First determination of reliable radii for Pluto and Charon (Tholen & Buie).
- 1986 First radiothermal measurements of Pluto's surface temperature (Altenhoff et al.).
- 1987 Water ice is discovered on Charon (Buie et al.; Marcialis et al.).
- 1988 Stellar occultation reveals Pluto's atmosphere (Elliot et al.).
- 1988 Formation of informal group interested in Pluto mission.
- 1989 Pluto at perihelion; *Voyager 2* flyby of Neptune and Triton.
- 1989 Mutual events produce evidence for polar caps on Pluto (Binzel).
- 1990 First NASA-sponsored Pluto mission study (*Pluto 350*) is completed.
- 1991 Publication of arguments that ice dwarf planets like Pluto were common in the early solar system (Stern).
- 1991 *Pluto 350* mission concept is replaced by *Mariner Mark II* mission concept.
- 1991 OPSWG is formed.
- 1992 Discovery of nitrogen and carbon monoxide ices on Pluto (Owen et al.).
- 1992 First of many Kuiper Belt objects is discovered (Jewitt & Luu).
- 1992 *Mariner Mark II* is replaced by *Pluto Express* mission concept.
- 1993 Evidence for large thermal differences across Pluto's surface (Stern, Weintraub & Festou; Jewitt).
- 1993 *Pluto Fast Flyby* enters Phase A development.
- 1994 Methane is detected in Pluto's atmosphere (L. Young et al.).
- 1995 *Pluto Fast Flyby* is canceled, replaced by *Pluto Express* as a proposed mission concept.
- 1995 Charon's nonzero lightcurve amplitude is discovered (Buie & Tholen).
- 1996 Hubble Space Telescope detects numerous surface features on Pluto (Stern, Buie & Trafton).
- 1997 Clyde Tombaugh dies. 600 km diameter ice dwarfs discovered (Luu et al.).

SUGGESTED READINGS

General Articles

In *Scientific American, 143* (1930), various articles: "Planet X" (p. 20); "How Pluto's orbit was figured out" (p. 364); "More about Pluto" (p. 446).

"The discovery of Pluto" by H. Giclas. *Icarus, 44,* 7–11 (1980).

"The discovery of Pluto: Some generally unknown aspects of the story" by C. W. Tombaugh. *Mercury, 15,* pp. 66–72 (Part I) and pp. 98–102 (Part II) (1986).

"Discovering Pluto's atmosphere" by J. K. Beatty and A. Killian. *Sky and Telescope, 76,* pp. 624–627 (1988).

"Pluto: The planet that came in from the cold" by N. Henbest. *New Scientist, 122*(1662), pp. 39–44 (1989).

"Pluto" by R. P. Binzel. *Scientific American, 262*(6), pp. 50–58 (1990).

"Where has Pluto's family gone?" by A. Stern. *Astronomy, 20*(9), pp. 40–47 (1992).

"The last world" by D. Sobel. *Discover, 14*(5), pp. 68–76 (1993).

"At the edge of night: Pluto and Charon" by R. Burnham. *Astronomy, 22*(1), pp. 40–47 (1994).

"To Pluto by way of a postage stamp" by R. L. Staehle, R. J. Terrile, and S. S. Weinstein. *The Planetary Report, 14*(5), pp. 4–11 (1994).

"Hot science on a cryogenic world" by A. Stern. *The Planetary Report, 14*(5), p. 8 (1994).

"Trip to Pluto" by R. L. Staehle and others. *Spaceflight, 36,* pp. 101–104 (Part I) and pp. 140–143 (Part II) (1994).

"Pluto and company" by P. Spence, I. Seymour, P. Bond, and S. Clark (four-article feature). *Astronomy Now, 9*(5), pp. 37–47 (1995).

"The discovery of Charon" by J. Christy. In *Pluto and Charon* (S. A. Stern & D. J. Tholen, eds.). University of Arizona Press, 1997.

"Clyde Tombaugh" (obituary), by Alan Stern, *Nature.*

Nontechnical Books

Scientific Monthly by W. Putnam and V. Slipher, *34,* 5–21 (1932).

The Search for Planet X by Tony Simon. Basic Books (1962).

Planets X and Pluto by W. G. Hoyt. University of Arizona Press (1980).

Out of the Darkness: The planet Pluto by C. Tombaugh and P. Moore. Stackpole Books (USA) and Lutterworth Press (UK) (1980).

The Planet Pluto by A. J. Whyte. Pergamon Press (1980).

The New Solar System (J. K. Beatty & A. Chaikin, eds.) Cambridge University Press and Sky Publishing Corporation (1990) (3rd ed.).

Clyde Tombaugh: Discoverer of Planet Pluto by D. H. Levy. University of Arizona Press (1991).

Exploring Planetary Worlds by D. Morrison. Scientific American Library (1993).

Technical Articles and Books

"The Discovery of Charon: Happy accident or timely find?" by R. M. Marcialis. *Journal of the British Astronomical Association, 99*(1), pp. 27–29 (1989).

"The Pluto–Charon System" by A. Stern. *Annual Reviews of Astronomy and Astrophysics, 30,* pp. 185–233 (1992).

"Solar system evolution: A New Perspective" by Stuart Ross Taylor. Cambridge University Press (1992).

"The Pluto reconnaissance fly-by mission" by A. Stern. *EOS, 74*(73), pp. 76–78 (1993).

Pluto and Charon (S. A. Stern & D. J. Tholen, eds.). University of Arizona Press, 1997.

The Origin of the Earth. (H. E. Newson, ed.) Oxford University Press (1990).

Information Available via the World Wide Web

The following sites all contain links to other relevant sites. (The availability and content of WWW sites are subject to continuous change by the individuals and organizations maintaining them. The following were accessible when the list was compiled in mid-1997.)

Marc Buie's Homepage (at Lowell Observatory)
http://www.lowell.edu/users/buie/pluto/pluto.html

Los Alamos National Laboratory
http://bang.lanl.gov/solarsys/pluto.htm#intro

University of Colorado/Fran Bagenal's Homepage
http://dosxx.colorado.edu/plutohome.html

Pluto Express Homepage at JPL
http://www.jpl.nasa.gov/pluto/

SEDS (Students for the Exploration and Development of Space)
http://www.seds.org/billa/tnp/pluto.html

INDEX

973-763 9120

973-992-1451

Mary Corvelli

rte 10

Livingston

right So Livingston Ave

East Cedar St

make left

4 the left Kearny Dr

#14

---

admitted

program will

Keith

Aug 30 4pm